Broxburn Shale
The Rise and Fall of an Industry

Broxburn Shale
The Rise and Fall of an Industry

Peter Caldwell

The Grimsay Press

The Grimsay Press
an imprint of
Zeticula
57 St Vincent Crescent
Glasgow
G3 8NQ
Scotland.

http://www.thegrimsaypress.co.uk
admin@thegrimsaypress.co.uk
First published 2010

Copyright © Peter Caldwell 2010

Every effort has been made to trace possible copyright holders and to obtain their permission for the use of any copyright material. The publishers will gladly receive information enabling them to rectify any error or omission for subsequent editions.

ISBN-10 1 84530 067 X Paperback
ISBN-13 978 1 84530 067 8 Paperback

All rights reserved. No part of this publication may be reproduced, stored in a retrieval system, or transmitted in any form or by any means, electronic, mechanical, photocopying, recording or otherwise, without the prior permission of the publishers.

Dedicated to the memory of those individuals
who lost their lives
in the Broxburn shale mines and oil works, and
to the families who endured hardship
when the local shale industry was forced to close.

Introduction

Our modern world is dominated by oil.

Many of the policies of national governments are based on the desire for its acquisition. Regimes have been overthrown in order to gain control of this valuable commodity. And it wasn't that much different one hundred years ago - oil was just as essential.

This is the story of the great shale oil industry which once flourished in the village of Broxburn, West Lothian, telling how Broxburn Oil Works extracted oil from the shale and created its many products.

The volume of shale excavated from the mines by the Broxburn Oil Company and the fees paid both to Lord Cardross — the Earl of Buchan — and to local estates for transporting shale across their land is given in detail.

Some personal background is added to those several individuals who appear otherwise to exist only as names associated with the early oil works, giving substance to these men.

The book also charts the rise and fall of local oil companies, and lists the fatal accidents which occurred in the shale mines and in the Albyn and Broxburn oil works.

Finally, it describes the sequence of events which led eventually — despite the continued importance of oil — to the closure of the shale mines and the demolition of Broxburn Oil Works.

No book of this kind can be attempted without the help and support of many people. Over a period of years, memories fade, and connections are lost.

For help with photographs, I should like to thank Peter Godfrey (Lothian Photographers), Edinburgh, National Monuments Record of Scotland, Edinburgh, and the RAF/Ministry of Defence.

In addition, my thanks are owed to the staff at INEOS, Grangemouth; National Archives of Scotland (HM General

Register House, West Register House), Edinburgh; University of Warwick BP Archives, Warwick; National Library of Scotland, (Map Library), Edinburgh; North of England Institute of Mining and Engineering, Newcastle Upon Tyne; West Lothian Local History Dept., Blackburn; Mining Records Office, Burton on Trent; and the British Geological Survey, Edinburgh.

Finally, I should be pleased to hear, through the publisher, from any one who has additional information or photographs which may add to the story.

Peter Caldwell
West Lothian
December 2009

Contents

Introduction	*vii*
List of Photographs	*xi*
Robert Bell	1
Broxburn Shale Oil Company Limited	6
Shale Mines and Oil Works	12
Broxburn Shale Mines and Pits	16
Oil Works at Broxburn	17
Glasgow Oil Company Broxburn Limited	21
Broxburn Oil Company, 1877-1901	28
Retorting and Refining at Broxburn	38
Broxburn Oil Company, 1903-1919	43
The Scottish Oil Agency & Scottish Oils	47
Scottish Oils Limited	52
Broxburn Oil Company, 1920-1925	53
A Question of Profits	75
The Anglo-Persian Oil Company and British Petroleum	79
Official Closure, 1927	81
A New Phase	83
Conclusion	86
Appendices	89
Norman Macfarlane Henderson	90
The Broxburn Oil Company's Mines to the Dunnet Shale	93
Description of the Broxburn Works	99
Shale Oil Works Chronology	104
Early Management and Workforce of Broxburn Oil Company	110
Broxburn Oil Company Shale Output, 1877 - 1925	111
Broxburn Oil Company Wayleaves, 1906 - 1925	125
Broxburn Oil Company Dividends, 1877 - 1910	131
The Shale Mine Years	132
The Shale Bings	133
Local Names for the Shale	134
Shale Worked by Broxburn Oil Company Limited	135
Glossary	*136*
Index	*139*

List of Photographs

The old coal pits south of Hayscraigs and Pyothall (map of 1855)	2
Robert Bell's mining area from north of Hayscraigs towards Pyothall	3
Bell's miners' rows at Holygate on a map of 1897	5
Broxburn Lodge (late 19th century)	8
Almondell House	9
Clifton Hall	10
The Fever Hospital (map of 1895)	11
Horizontal Retort	15
Poole Hall	18
Spylaw House	18
Furnival's Inn	19
Certificate of Incorporation of The Broxburn Oil Company	27
Certificate of Incorporation of The Scottish Oil Agency	47
Certificate of Registration .. confirming Alteration of Objects	48
Certificate of Incorporation of Scottish Oils	50
Certificate, that Scottish Oils is entitled to commence business	51
Two views of Broxburn Oil Works around 1900	56
Albyn Oil Works c. 1900	57
Miners in Hayscraigs Shale Mine, c. 1910.	58
Man at the top shot boring on an incline, c. 1900	59
Railway within Broxburn Oil Works, c. 1910	60
Broxburn Oil Company Locomotive, c.1920	60
The Iron Bridge over west Main street c. 1905	61
Boring a shot-hole in which to place the charge with compressed air drill, c.1910	62
Diesel locomotive pulling hutches loaded with shale, c.1915	62
Broxburn Oil Works, c. 1910	63
Broxburn Oil Works from Broxburn Gas Works, c.1910	64
Broxburn Candle House, in the 1930s	64
Advertisement for Broxburn Paraffin Candles	65
Greendykes Road, c. 1908	66
Greendykes Road, 1916	67
Greendykes Road, c. 1925	67

Broxburn Oil Company's houses at Greendykes, 1960	68
Stewartfield Rows, c.1935	69
The old Oil Company houses	69
Houses in the Stewartfield Rows in the 1960s	70
Broxburn shale bings in 1982	71
Albyn tip, now gone, beside Bridge No 29 in 1996	72
Bridge No 28, where the Broxburn shale industry began	72
An aerial view of the Broxburn shale bings, in 1947	73
Entrances to the Broxburn shale mines and pits	74
The Henderson Retort	91
Plan of workings of Dunnet Shale Mine	93
Rock Strata tunnelled through to the Dunnet Shale Mine	94
Electrical Power-House	95
Endless-Rope Haulage-Gear at the Dunnet Shale Mine	96
Dunnet Shale Mine beside Albyn Crude Oil Works, 1910	98

Robert Bell

The Broxburn shale-mining industry began in the early 1860s because of one man, Robert Bell, who was a coal master (owner or lessee of a coalfield). Bell was granted a lease in 1858 by the Earl of Buchan to search for coal and ironstone at Broxburn. It had been known for some time that coal was to be found there. Coal pits had been worked in Broxburn before, one at Pyothall (Thomson's Coal Pit) and one at Hayscraigs.

In fact, in the late 1850s, prior to Bell's arrival, there was already a coal master in Broxburn, a man named John Cunningham. In 1860, Cunningham the coal master was still there, but in 1861 Robert Bell, coal master, is listed. It may be that Bell simply took over coal mining operations in Broxburn from Cunningham. If that was so, it was bad timing for John Cunningham.

Robert Bell discovered shale when coal was being mined in Broxburn. His original shale mining operations took place in an area between Hayscraigs and Pyothall.

Bell's shale discovery laid the foundation of the great shale-oil industry. He is credited with being the first man in Scotland to distil oil from shale.

Robert Bell's original coal and shale mining operations (only the shale mine workings are shown on the map) began north of Hayscraigs Farm, and incorporated the old coal pit to the south-east of Hayscraigs.

John Cunningham might have opened up the old coal mine, which, like that near Pyothall, had been abandoned for about twenty years. Robert Bell fortuitously took over the coal operation just before shale was discovered. Bell's first shale mine was in operation for about thirteen years.

The system used in mining coal was the longwall method. This consisted of miners employed at a long working face in which a coal seam was exposed along its length and removed layer by layer. As the men worked along the seam, wooden props supported their workspace. Props also supported the roofs of the roadways which led to the coal face. As the miners

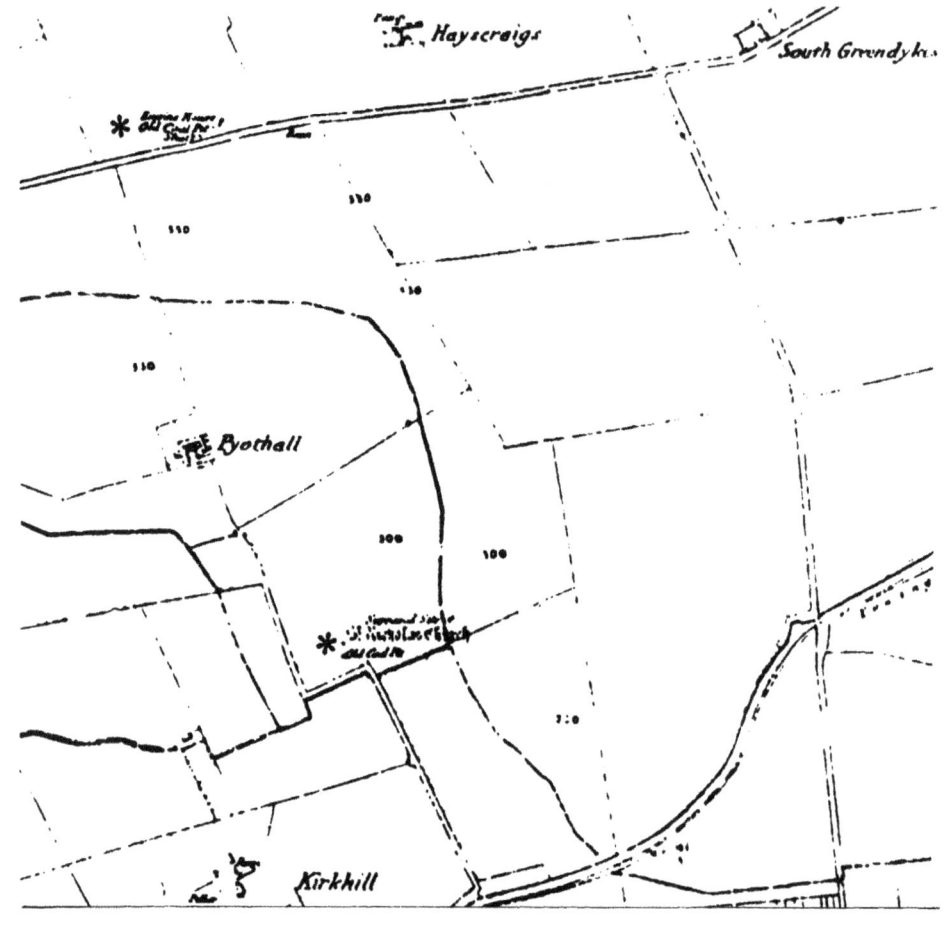

The old coal pits south of Hayscraigs and Pyothall. (Map of 1855)

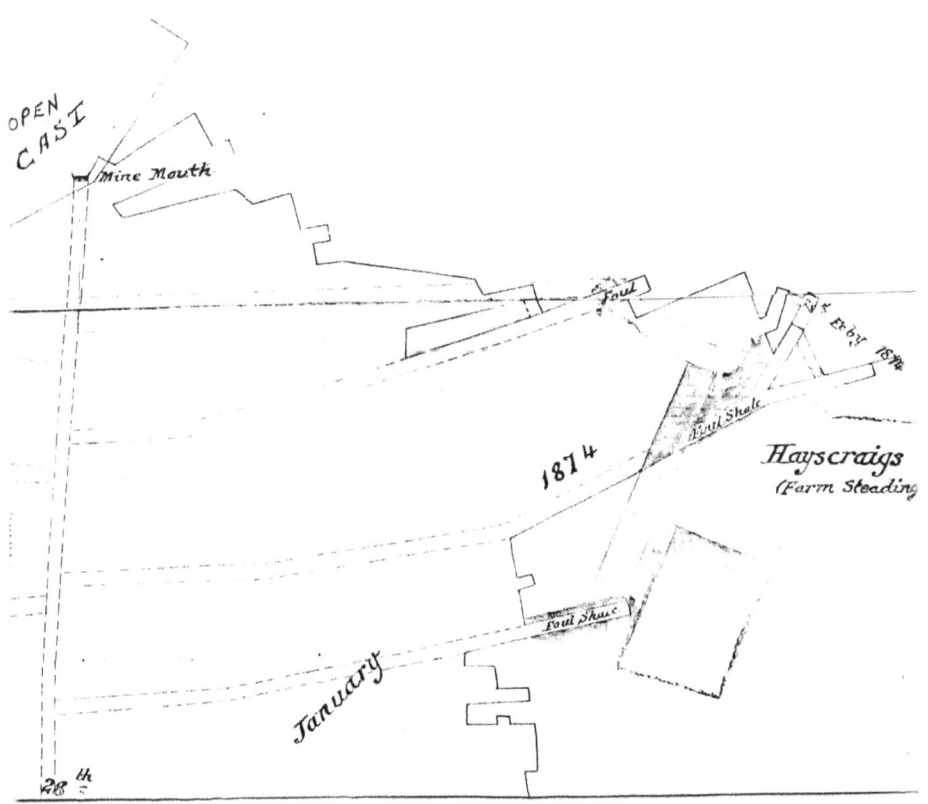

Robert Bell's mining area from north of Hayscraigs towards Pyothall. The mine entrance was near the old farm of North Greendykes after which the mine was named.

advanced along the long wall the supporting props were moved forward with them and the area behind, having been worked out, was allowed to cave in.

The longwall method of coal mining was used to mine the shale in the North Greendykes Shale Mine, and was used in all the early shale mining operations.

Son of a farmer, Robert Bell was born in Wishaw in 1824. He started in business as a wood merchant. Still in that business, in 1851 he leased a coalfield on the Wishaw Estate and became one of the leading coal masters there, establishing the Wishaw Iron Works in 1859. During his coal operations in Broxburn, sometime in 1861 deposits of opencast shale were discovered.

Broxburn was a rural community where most of the local people worked on the farms, so Bell would almost certainly have advertised for miners in national newspapers.

Workers and their families needed homes to live in, and the advent of shale operations sounded the death knell for Holygate Farm. The tenure of the farm lease may have been bought out by Robert Bell, because in 1862 he had six rows of houses (room and kitchen) built for his miners and their families on part of the farm. These miners' rows, the first to be built in Broxburn, were close to Holygate farmhouse, and occupied the area where the houses of Holygate Place now stand.

The setting up of mining operations began, and another shale mine and pit near his North Greendykes mine was opened up. The construction of an oil works was initiated, the Broxburn Shale Oil Company being formed in 1862.

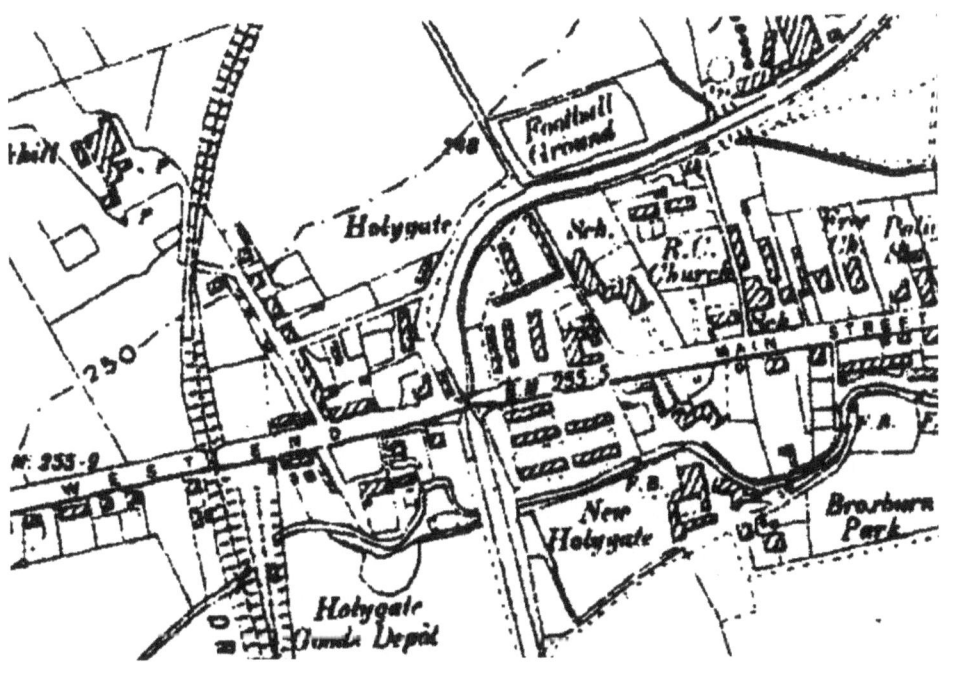

*Bell's miners' rows at Holygate on a map of 1897.
Old Holygate Farmhouse is the **C** shape just to the right of the three rows
of houses.*

Broxburn Shale Oil Company Limited

Incorporated 27 March, 1862
Registered Office: 36 Renfield Street, Glasgow

Nominal Capital:
£20,000 divided into 100 shares of £200 each

Company Subscribers
William Brown, Timber Merchant, 23 Corn Street, Glasgow (20 shares)
Alexander Robertson, Commission Merchant, 146 Buchanan Street, Glasgow (3 shares)
Alexander Dick, Writer, 153 Queen Street, Glasgow (2 shares)
William Henshaw, Manufacturer, 17 John Street, Glasgow (10 shares)
William Henshaw, Jun., Manufacturer, 17 John Street, Glasgow (2 shares)
Robert Faulds, Jun., Merchant, 30 Parson Street, Glasgow (20 shares)
William Stephen, Gentleman, Viewfield, Campsie Junction (4 shares)

14 September, 1864
The registered Office moved to 125 Buchanan Street, Glasgow

Date of Liquidation - 14 September, 1864

Although Robert Bell was the central figure in the Broxburn shale oil business from the early 1860s onwards, he remained a coal master also, still working the coal at his mine at Hayscraigs. The manager of Bell's Broxburn Colliery up to 1867 was Alex Scott.

Robert Bell also had a brickworks - Newbigging Brick Works - near the colliery and shale mines.

Robert Bell married Agnes Dalrymple on the 24 March, 1869, in Braddon, Isle of Man.

Their children were
    John Dalrymple born 1869, Edinburgh
    Letitia Dalrymple born 1871, Edinburgh
    Janet Morton born 1872, Edinburgh / Uphall *
    Edith Margaret born 1874, Edinburgh / Uphall *
    William Dalrymple born 1877, Edinburgh
    Ruby Dalrymple born 1878, Edinburgh
    Robert Morton born 1879, Uphall
    Walter Dalrymple born 1880, Uphall
        Births registered in both places

Soon after his marriage, as well as his shale oil interests, coal mine, and brickworks, Robert Bell became a farmer - of Greendykes Farm.

Age 45 in 1871, he still lived in Broxburn Lodge. His wife is not listed with him, no doubt because she was in Edinburgh awaiting the birth of her second child. The census lists him as - Coal Master, Shale Oil Manufacturer and Farmer, employing 400 men and 30 boys.

In 1872, Robert Bell purchased the Holmes Estate where, in 1884, the Holmes Oil Works were constructed. Bell became a director, and the major shareholder in the Holmes Oil Company Limited.

In the 1870s, Robert Bell had been threatened with legal action because the streams near his oil works were being polluted with water discharged from them. He decided to pump this discharge up to the top of shale waste hoping it would evaporate. Instead, it gradually soaked right through the shale bing and oozed out the bottom edges. To his surprise, Bell noticed that the grass growing around this discharge was

Broxburn Lodge (late 19th century) — the part known as the Yellow House

much more luxuriant than the grass growing further away, so he decided to have the offending liquid analysed. The 'waste' was found to contain ammonium nitrate, a very valuable fertilizer.

This business grew so much that it became difficult for one individual to control, and it was one of the reasons why the Broxburn Oil Company Limited was formed in 1877 by Robert Bell.

In 1878, the Bell family moved to Almondell House, where they lived until 1884.

Almondell House

In 1884, Robert Bell bought the mansion-house and estate of Clifton Hall, between Broxburn and Newbridge, from Sir James Gibson Maitland for £50,000. The family resided here until 1894. During this same period,1882-1894, he also had a city residence at 8 Eglinton Crescent, Edinburgh.

Clifton Hall

For many years Robert Bell was a member of the Uphall Parochial Board and of the School Board. He was also a Justice of the Peace, and a County Councillor - when the position was unpaid. He appears to have been held in high regard by the local community.

He had a Public Hall, with town clock, built in 1872. The clock was preserved when the building was demolished in the 1960s. It is now in the Strathbrock Centre.

In 1881, Mr and Mrs Bell had a fully equipped Fever Hospital built and gifted to Broxburn. This was for the cure and prevention of contagious fevers. The well-known scarlet fever was one, from which you could become deaf, go blind, or die. There was also rheumatic fever and typhus fever, both of which could be fatal. The Fever Hospital stood at the present Newhouses Road.

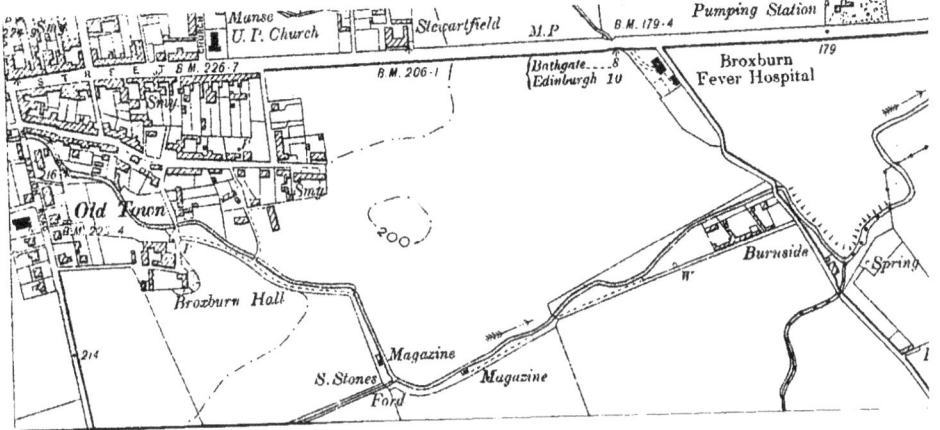

The Fever Hospital (map of 1895).

Agnes Bell died in childbirth (peritonitis) in 1883 at 8 Eglinton Crescent. Robert Bell died at Clifton Hall on 30 May 1894, of pleurisy and pneumonia.

Shale Mines and Oil Works

Shale is the result of clay having being compacted by pressure throughout vast geological time periods. Over many millions of years it has been laid down, one stratum on top of another, eventually forming through heat and pressure a hard rock such as coal or a less firm rock such as shale, depending on, amongst other things, the amount of organic material contained within it.

Shale is a finely-layered, soft rock, that splits easily, but not all shale contains oil. There are two different types, ordinary shale and oil shale.

The shale could be extracted with picks from the harder rock surrounding it, but more often it was mined by blasting.

Owing to geological movement, the depth of shale seams was not uniform, and a seam could be at a different depth in a neighbouring area. Seams could vary in places because they may have been pushed upwards closer to the surface, or plunged downwards, by movements of the strata underground. Seams could also be broken due to faulting, the remainder of the seam being above or below the original part being worked. The rest of the seam would have to be searched for, but might never be located.

Oil shale is dark brown or almost black, and contains long-decayed vegetable matter composed of algae and other plant material, and this has formed a substance called kerogen.

Oil shale sometimes contained enough kerogen to enable it to burn without any processing, and it was often known as 'the rock that burns.'

Over aeons of time, the geological actions on ancient organic matter could result in oil or oil shale, the basic difference being that the pressure and heat which resulted in the production of oil was greater than that which produced oil shale.

Although the kerogen in the shale has not yet been changed by the friction of the rocks into oil, mineral oil can be distilled from it. Heating the rock to high temperatures vaporises the kerogen, and this vapour condenses to form a slow-flowing oil, a petroleum-like liquid called shale oil.

Shale oil was a crude oil, and in itself was not of much use, so the building of an oil works (crude oil works) to treat the oil obtained by distillation was necessary. This was a paraffin oil works for the manufacture of paraffin and paraffin wax from the extracted shale oil.

Paraffin was the major oil product in the 19th and early 20th centuries. Paraffin lamps were ubiquitous. Paraffin was also used in paraffin stoves for heating and cooking. Paraffin wax was used to make candles. The paraffin lamps, stoves, and candles would have been in almost every house in the country.

Over several decades many mines and pits were excavated on the Buchan Estate by blasting then digging with picks and shovels. Some of the shale was near the surface, and this was worked as opencast. Many of the mines and pits were named after local farms.

Mines sloped downwards to allow access to shale, the shafts (roadways) being about 12 feet wide and could be up to 10 feet in height. The men walked down the inclined shaft to their place of work, or rode on a bogie train. Entrances to the mines were built with bricks which, from the outside, made the shafts similar in appearance to railway tunnels.

Pits were vertical shafts, usually round, lined with bricks, and were sunk to reach the deeper shale seams. They were normally about 15 feet in diameter. Men reached the pit bottom via cages, two of which travelled up and down each shaft.

At the bottom of a pit or mine shaft - formed by blasting with gunpowder then using picks and shovels - a network of tunnels at right angles to each other was created stretching outwards. When mining, large square pillars of shale were left in place by the men to support the roof of the mine. The shale was obtained by blasting. Holes were bored by hand into the shale, a charge of compressed gunpowder was pushed into the hole using a copper covered tool to avoid a spark, then the charge was set off. A drawer loaded the shale into a hutch and pushed it to the main haulage road (which could be up to ½ a mile from him) where the hutches, each holding about 1 ton of shale, were drawn mechanically along the road by a rope haulage system to the pit bottom. They were then raised to the surface in a steel cage by means of steam driven winding gear. When a particular area was cleared of shale, working inwards from the farthest part of the mine, each pillar of shale was then worked out, leaving the roof at this vicinity to subside. This was not done in areas which lay beneath houses, rivers, etc.

At the surface, the shale-laden hutches were drawn along a narrow railway to the retorting plant where the shale was put through steel rollers which broke it into pieces.

It was then fed into retorts and heated, turning the kerogen contained in the shale into vapour which then passed through pipes to condensers where, when cooled, it formed shale oil and ammonia water. The oil and ammonia water were then separated and passed through pipes to different tanks.

Gases containing spirit and some ammonia which did not condense in the condensers were piped to 'scrubbers' (steel towers) where the products were extracted.

The dark green crude oil was pumped into tanks, and from there it was piped into what were known as continuous stills to be refined. These stills were connected to each other, each being kept at a different temperature. The crude oil passed through each in turn, and a different product was extracted because of the variance of the heat within each still.

Just some of the many products of shale oil refining were naphtha (an inflammable oil later used as petrol), kerosene (paraffin); lubricating oil; wax for making candles, matches, electrical insulating, waterproof papers, etc; coal gas for heating, lighting and cooking; ammonium nitrate for fertilizers, and to make explosives; solvents for paint, and rubber manufacturing; sulphuric acid for use in the iron and steel making industry; and petroleum jelly.

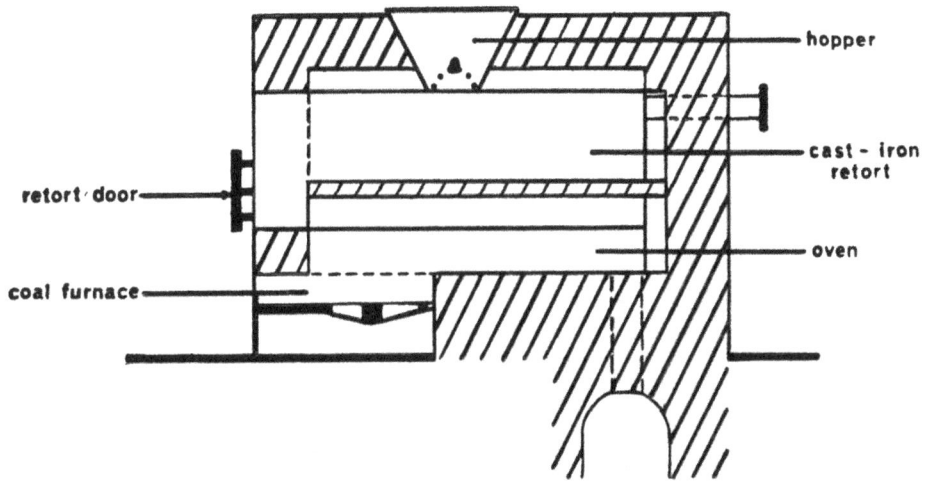

Horizontal Retort

Broxburn Shale Mines and Pits

[Shale names as listed in Mining Authority Records]

Albyn Mine - Broxburn Shale, Curly Shale.

Dunnet Mine - Dunnet Shale.

Greendykes North Pit - The type of shale is not specified on the Mining Authority Records. [Opencast Shale]
Greendykes South Pit - Broxburn Shale, Grey Shale, Curly Shale.

Hayscraigs Mine - Broxburn Shale, Grey Shale, Curly Shale.

Hut Mine - Broxburn Shale, Grey Shale, Curly Shale.
Hut Pit - Broxburn Shale, Curly Shale.

Pyothall No 5 Pit (*also known as* Hayscraigs Pit) Broxburn Shale, Grey Shale, Curly Shale.

Stewartfield No 1 Mine - Broxburn Shale, Grey Shale, Curly Shale, Fells Shale, Dunnet Shale.
Stewartfield No 1 Pit - Broxburn Shale, Curly Shale.
Stewartfield No 2 Pit - Broxburn Shale, Curly Shale.
Stewartfield No 3 Pit - Broxburn Shale, Curly Shale.
Stewartfield No 4 Pit - Broxburn Shale, Curly Shale.

Oil Works at Broxburn

In 1862, Robert Bell had Faulds of Glasgow construct an oil works with 36 horizontal retorts for him at Greendykes, just north of canal bridge No 28. Faulds extracted the crude oil for the Broxburn Shale Oil Company Limited.

Dr James Steele, another Wishaw man, set up an oil works, also on the north side of the canal, to the west of Bell's works, near Greendykes Road.

Works in 1862
Oil works of Broxburn Shale Oil Company, Greendykes (R. Bell)
Oil works of James Steele, Greendykes

In 1863, Robert Bell had an oil works with 100 horizontal retorts built for himself between Stewartfield Farm and the canal.

Works in 1863
Oil works of Broxburn Shale Oil Company, Greendykes (R. Bell)
Oil works of James Steele, Greendykes
Oil works of Robert Bell, Stewartfield

The Broxburn Shale Oil Company Limited was wound up in 1864, its assets (the oil works) being sold to Edward Fernie of Poole Hall, Nantwich, Cheshire, formerly of St David's Oil Works, Saltney, Cheshire, a coal oil refinery owned by the Flintshire Oil and Cannel Company.

He added a further 32 vertical retorts to those already there. Horizontal retorts were said to produce more paraffin, while vertical retorts were said to produce a better quality crude oil. Both types were used together for many years.

Thomas Hutchison of Spylaw House, Colinton, Edinburgh, built a small oil refinery on the west side of Greendykes Road, opposite the oil works of Dr James Steele. Hutchison refined oil for Robert Bell.

Poole Hall

Spylaw House

Works in 1864
Oil works of Edward Fernie, Greendykes
Oil works of James Steele, Greendykes
Oil works of Robert Bell, Stewartfield
Oil refinery of Thomas Hutchison, Greendykes

In 1865, the oil works of Edward Fernie were taken over by Ebenezer Waugh Fernie.

Brought up in London, he was actually born (1815) in Geissen, Darmstadt, Hesse, Germany. Before he came to Broxburn he and his family lived in Furnival's Inn, Holborn, London, where he was described as a merchant. He had been previously listed as being the proprietor of a mine in Devon. Charles Dickens lived in Furnival's Inn 1834-1836.

Ebenezer Waugh Fernie lived in Almondell House while engaged in the shale oil business, dying there on 29th May 1869.

Furnival's Inn

Robert Bell had another oil works built with 100 horizontal retorts at Greendykes.

James Steele had another shale oil works built, this time at Stewartfield.

James Vallance, a chemist from Paisley, had an oil refinery built in Almondfield.

This refinery stood just east of the Arches on the south side of the present main road. In the field directly opposite, on the north side of the road, there was a shale mine (Newliston Shale Mine).

Works in 1865
Oil works of Ebenezer Waugh Fernie, Greendykes
Oil works of James Steele, Greendykes
Oil works of Robert Bell, Stewartfield
Oil refinery of Thomas Hutchison, Greendykes
Oil works of Robert Bell, Greendykes
Oil works of James Steele, Stewartfield
Oil refinery of James Vallance, the Arches

While engaged in the running of his shale oil business, Robert Bell resided in Broxburn Lodge, on the opposite side of the street from his miners' rows at Holygate.

In 1866, John Edgar Poynter, a manufacturing chemist of 72 Great Clyde Street, Glasgow, took over James Steele's shale oil works at Greendykes. He named them Buchan Oil Works.

Robert Bell formed a new company, Glasgow Oil Company Broxburn Limited, and this company took over the oil works of Ebenezer Waugh Fernie (died at Almondell House of paralysis on 29 May 1869, age 54).

E.W. Fernie's were the original works of Robert Bell's Broxburn Shale Oil Company Limited, which Edward Fernie had enlarged. Bell named the new works Albyn Shale Oil Works.

In 1866, James Miller, a banker from Wishaw, and, like Robert Bell, a former director of Glasgow Oil Company Broxburn Limited, set up his own oil works at Greendykes.

Glasgow Oil Company Broxburn Limited

Incorporated 27 June, 1866

Registered Office: 48 Dundas Street, Glasgow

Nominal Capital: £25,000 divided into 500 shares of £50 each

First Directors: Robert Bell, Walter McLellan, James Miller

Company Subscribers
Robert Bell, Coal Master, Broxburn (125 shares)
Walter McLellan, Jun., Iron Merchant, 129 Irongate Street, Glasgow (125 shares)
James Hamilton, Coal Master, Queens Park, Glasgow (20 shares)
James Miller, Banker, Wishaw (50 shares)
Robert Faulds, Waggon Builder, 61 Parliamentary Road, Glasgow (50 shares)
Benjamin Conner, Engineer, Caledonian Railway, St Rollox, Glasgow (10 shares)
William Scott, Jun., Coal Master, Wishaw (50 shares)
James Lythgoe, Accountant, North British Railway Company, Joppa (20 shares)
William Hurst, Engineer, Edinburgh (50 shares)
Date of Liquidation - 27 April, 1875

```
Fatal Accident in Shale Mine

22 Jan 1866 - Broxburn No 2 Pit.
(Robert Bell)

   An accident occurred at Broxburn
Shale Works, of which Robert Bell is
lessee, about midday on Monday, by
which a miner named John Shields, about
40 years of age, was instantaneously
killed. It appears that Shields was
at his usual employment in No 2 shale
pit and that while engaged in holing
in a face of shale, a large mass
became detached, and crushed him so
severely that he almost immediately
expired. (Hamilton Advertiser, 27
Jan 1866)
```

Works in 1866
Albyn Oil Works of the Glasgow Oil Company Broxburn, Greendykes
Buchan Oil Works of John Poynter, Greendykes
Oil works of Robert Bell, Stewartfield
Oil refinery of Thomas Hutchison, Greendykes
Oil works of Robert Bell, Greendykes
Oil works of James Steele, Stewartfield
Oil works of James Miller, Greendykes
Oil refinery of James Vallance, the Arches

In 1868, the refinery of Thomas Hutchison and the oil works of James Steele were abandoned.

Works in 1868
Albyn Oil Works of the Glasgow Oil Company Broxburn, Greendykes
Buchan Oil Works of John Poynter, Greendykes
Oil works of Robert Bell, Stewartfield
Oil works of Robert Bell, Greendykes
Oil works of James Miller, Greendykes
Oil refinery of James Vallance, the Arches

In 1869, the operating of the oil works of James Miller was undertaken by the Glasgow Oil Company Broxburn Ltd. The works were still privately owned by James Miller.

James Liddell of 301 Eglinton Street, Glasgow, set up an oil works on the west side of Greendykes Road, just north of the canal - more likely, he put the recently abandoned works of James Hutchison back into production.

The former oil works of James Steele lay unused.

Works in 1869
Albyn Oil Works of the Glasgow Oil Company Broxburn, Greendykes
Buchan Oil Works of John Poynter, Greendykes
Oil works of Robert Bell, Stewartfield
Oil works of Robert Bell, Greendykes
Oil works of James Miller, Greendykes
Oil works of James Liddell, Greendykes
Oil refinery of James Vallance, the Arches

From 1871, the oil refinery of James Vallance is no longer mentioned in records. It may have been affected by the increased importation of American kerosene (burning oil). The price of paraffin in the mid 1860s was around 1/6d per gallon. This fell to about 10d per gallon in the 1870s because of cheaper imported burning oil.

Works in 1871
Albyn Oil Works of the Glasgow Oil Company Broxburn, Greendykes
Buchan Oil Works of John Poynter, Greendykes
Oil works of Robert Bell, Stewartfield
Oil works of Robert Bell, Greendykes
Oil works of James Miller, Greendykes
Oil works of James Liddell, Greendykes

In 1872, Robert Bell set up a refinery to the west of Greendykes Road.

At Greendykes, the oil works of James Miller and John Poynter's Buchan Oil Works were unused.

The oil works of James Steele at Stewartfield lay abandoned.

Works in 1872
Albyn Oil Works of the Glasgow Oil Company Broxburn, Greendykes
Oil works of Robert Bell, Stewartfield
Oil works of Robert Bell, Greendykes
Oil works of James Liddell, Greendykes
Oil refinery of Robert Bell, Greendykes

By 1873, James Liddell was now James Liddell & Company, 6 Alison Street, Crosshill, Glasgow.

Buchan Oil Works of John Poynter at Greendykes, the oil works of James Miller at Greendykes, and that of James Steele at Stewartfield, lay unused.

In 1874 Robert Bell's North Greendykes Shale Mine closed. His coal mine there remained in production, as did the South Greendykes Shale Mine.

In April 1875, the Glasgow Oil Company Broxburn Ltd was wound up, Albyn Shale Oil Works reverting to Robert Bell. Buchan Oil Works, and the former oil works of James Miller and James Steele, lay abandoned.

Works in 1875
Albyn Oil Works of Robert Bell, Greendykes
Oil works of Robert Bell, Stewartfield
Oil works of Robert Bell, Greendykes
Oil works of James Liddell & Company, Greendykes
Oil refinery of Robert Bell, Greendykes

In 1876, the oil works of Robert Bell at Stewartfield were closed.

The oil works of Wishaw banker James Miller were back in operation.

The registered address of James Liddell & Company was now 37 Houston Street, Glasgow.

At Greendykes, John Poynter's Buchan Oil Works lay abandoned, as did the oil works at Stewartfield, once owned by James Steele.

```
Fatal Accident in Shale Mine

9 Feb 1876 - Pyothall. (Robert
Bell)
   David Beveridge, 50, pitheadman.
He neglected to shut the door after
taking off a tub, and on returning
with the 'empty' the cage was away
and he ran it into the shaft.
```

Works in 1876
Albyn Oil Works of Robert Bell, Greendykes
Oil works of Robert Bell, Greendykes
Oil works of James Liddell & Company, Greendykes
Oil refinery of Robert Bell, Greendykes
Oil works of James Miller, Greendykes

In 1877, George Simpson of 14 Maitland Street (now Shandwick Place), Edinburgh, decided to have an oil works built near East Mains, close to the Stewartfield shale mines.

```
Fatal Accident in Shale Mine

18 Aug 1877 - Broxburn. (Robert
Bell)
   J. Bryans, 24, brusher. Run over
on engine plane.
```

In 1877, Robert Bell formed the Broxburn Oil Company Limited. This new company bought up the oil works (except for one) not owned by Robert Bell, together with any mineral rights. The only one now privately owned was George Simpson's at East Mains. Did he refuse to sell?

During this major transformation period in the Broxburn shale oil industry, Robert Bell resided at 6 Grosvenor Street, Edinburgh from 1876 to 1878.

The former oil works of Robert Bell, James Liddell & Company, and James Miller, (all at Greendykes) would have been kept in production while a new, larger, Albyn Oil Works was built on its old site east of Greendykes Road. On the west side of Greendykes Road, the construction of a large, modern refinery was begun, so Bell's refinery must also have been kept in production until these two engineering projects were completed. The refinery was commonly listed as Broxburn Oil Works.

Works in 1877
Albyn Oil Works of Broxburn Oil Company, Greendykes
Oil works of Broxburn Oil Company, Greendykes
Oil works of Broxburn Oil Company, Greendykes
Oil refinery of Broxburn Oil Company, Greendykes
Oil works of George Simpson, East Mains

REGISTRATION N° 792

CERTIFICATE OF THE INCORPORATION OF A COMPANY.

I hereby Certify that

"The Broxburn Oil Company (Limited)"

was INCORPORATED under the Companies Acts *1862 and 1867* on the *Sixth* day of *November,* One Thousand *Eight Hundred and Seventy seven.*

Given under my hand at Edinburgh, this *Thirteenth* day of *December* One Thousand Nine Hundred *and Two.*

R R Macgregor
for Registrar of Joint-Stock Companies for Scotland

Companies Act, 1862, Sec. 174 (5).

Certificate of Incorporation of The Broxburn Oil Company

Broxburn Oil Company, 1877-1901

Broxburn Oil Company Limited
Incorporated 6 November, 1877
Registered Office - 28 Royal Exchange Square, Glasgow
Nominal Capital - £165,000 divided into 16,500 shares of £10 each
(Known) Directors:
William Kennedy, Robert Bell, John Waddell
Company Secretary:
John Nicholls King

Minute of Agreement
between
The Broxburn Oil Company (Limited)
and
Robert Bell

Dated 3rd and 5 February 1879

Relevant extracts from the above agreement [full of legal terminology] between Robert Bell and the recently-incorporated Broxburn Oil Company Limited:

'Thereas by Minute of Agreement - - - - - - - between the said Robert Bell and William Kennedy, oil Merchant in Glasgow, on behalf of the said Broxburn Oil Company (Limited), the said Robert Bell inter alia agreed to sell to the said Company, and the said William Kennedy, as acting on behalf of the said Company, agreed to purchase all his, the said Robert Bell's, rights and interest as tenant under a lease - - - - - - - - of the Coal, Oil, Shale, Ironstone, Fireclay and common clay in portions of the lands and Estate of Strathbrock and Kirkhill in the County of Linlithgow belonging to Lord Cardross, as well as all his, the said Robert Bell's, rights and interest in the Subjects known as the Broxburn Shale, Oil, Coal and Brick Works respectively, situated at or near Broxburn - - - - - - - - - - and generally the whole works, Houses, Buildings, Mining and manufacturing, Plant, Machinery, Apparatus, Railways, Tramways and other fittings belonging to - - - - - - - - - the said Robert Bell - - - - - - - and that at the agreed on price - - - - - - of Seventy five thousand pounds Sterling - - - - - - - - - and that of the said price of Seventy five thousand pounds, Sixty five thousand pounds should be paid in Six thousand five hundred fully paid up shares of Ten pounds each of the Broxburn Oil Company (Limited) - - - - - - - which shares Four thousand five hundred should be ordinary shares and Two thousand deferred shares - - - - - - - - - - - - Therefore, we, The Broxburn Oil Company (Limited) - - - - - - - - and I Robert Bell - - - - - - - - agree to accept of the allotment of Six thousand five hundred fully paid up shares of Ten pounds each of the said Company, of which shares Four thousand five hundred shall be fully paid up ordinary shares of Ten pounds each and numbered One to Four thousand five hundred inclusive, and Two thousand fully paid up deferred shares of Ten pounds each and numbered B One to B Two thousand inclusive - - - - - - - - - - - - - - - - - and these presents are sealed with the seal of the Broxburn Oil Company (Limited) and subscribed by John Waddell, Contractor, Edinburgh, and the said William Kennedy, two of the Directors, and by John Nicholls King, the Secretary of the said Broxburn Oil Company (Limited) - - - - - - - before these witnesses - - - - -'

    D. Forbes Witness
    John Waddell Director
    Geo. Ferguson Witness
    William Kennedy Director
    Wm. Mitchell witness
    John N. King Secretary
    Robert Bell

    "Reproduced from the BP Archive"

Lord Cardross

David Stuart Erskine became Earl of Buchan and Lord Cardross in 1857. It was from him that Robert Bell obtained his lease to search for minerals at Broxburn. However, things had changed.

In 1872, the Earl of Buchan made over his Linlithgowshire estates to his eldest son, Shipley Gordon Stuart Erskine, who had assumed his father's lesser title of Lord Cardross in 1871 when he became twenty-one years old.

The lands were transferred to his son in return for a yearly payment of £500 to the Earl of Buchan who seems to have spent most of his money on horses - strangely, he rode as a jockey. He was made bankrupt for a debt of £388 in 1894.

So, although Lord Cardross did not become the 14th Earl of Buchan till 1898, he, not the Earl of Buchan, owned the lands at Broxburn when Broxburn Oil Company Limited was formed in 1877.

William Kennedy

Managing Director of Broxburn Oil Company. Born Douglas, Lanarkshire, in 1836, he was only 41 when he was appointed. He did not stay locally, but lived in Braemar House, Nithsdale Road, Govan. He married Margaret Law from Linlithgow in Hutchesontown, Glasgow, in 1860. Three of their five children were born in Glasgow, two in Govan.

John Waddell

Director of Broxburn Oil Company, he lived at 4 Belford Park, Edinburgh. He was a Railway Contractor with 3,000 employees. Born New Monkland, Lanarkshire, in 1829, he married Margaret Donald in New Monkland in 1852. He must have moved to Bathgate during the period 1855-1856. Of their nine children, seven were born in Bathgate.

John Nicholls King

Secretary of Broxburn Oil Company, he lived at Mayfield House, New Monkland, Lanarkshire. Born in St Austell, Cornwall, in 1849, he married around 1878. His two known children were both born in New Monkland.

Robert Bell

Director of, and major shareholder in, Broxburn Oil Company. Chairman from 1887.

Broxburn Oil Company Ltd built over six hundred houses (room and kitchen) for their workmen in the shale mines and oil works, at Greendykes, Stewartfield, and New Holygate. These became known simply as miners' rows.

In 1878, the new Broxburn Refinery and Albyn Oil Works were completed. The building of the oil works was under the control of Norman Henderson who was appointed Works Manager. Albyn Oil Works obtained crude oil from the shale, and a pipeline over 400 yards long fed this oil to a tank at Broxburn Refinery for processing.

Works in 1878
Broxburn Refinery (Broxburn Oil Company), Greendykes
Albyn Oil Works (Broxburn Oil Company), Greendykes
Oil works of George Simpson, East Mains

In 1879, Broxburn Oil Company finally bought out George Simpson who built his oil works just before the 1877 takeover. Perhaps he had acquired inside information.

Works in 1879
Broxburn Refinery (Broxburn Oil Company), Greendykes
Albyn Oil Works (Broxburn Oil Company), Greendykes

In the 1880s, Broxburn Oil Company opened a Gas Works on the east side of Greendykes Road near Albyn Oil Works. This supplied the village of Broxburn.

```
Fatal Accident in Shale Mine

18 Mar 1880 - Broxburn.
  Henry Black, 55, miner. Fall of
shale while holing. Stooping.
```

Fatal Accident in Shale Mine

25 Jan 1881 - Broxburn. [Hayscraigs Mine]

William Armstrong, 29, miner. While riding up the mine, between two water barrels, his head struck the roof. It was against orders to ride on the incline.

Fatal Accident in Shale Mine

23 Aug 1882 - Broxburn. [Hayscraigs Mine]

Explosion of firedamp.

Francis Danks, 48, overman. John Steven, miner, 18.

John Imrie, miner, 24. John Neil, fireman, 26.

The report by the mine inspector blamed the men: 'Danks and Neil, without thought, went with naked lights into a place which they knew was unventilated. They all lived more than a week after the accident, and could explain how it happened.' Miners did not trust the safety-lamps, preferring their oil lamps. The much-vaunted Davy Safety Lamp emitted about one-eighth of the light given out by a candle, and, although there was a risk (methane explosions were more common in coal mines than shale mines), many shale miners chose to use naked-flame oil lamps instead.

Fatal Accident in Shale Mine

2 Mar 1883 - Broxburn.

David Bisset, 14, engine keeper. Crushed by gearing of pump which he was oiling while in motion.

The Registered Office of Broxburn Oil Company Limited moved to 18 Royal Exchange Square, Glasgow in 1885.

1886 - Broxburn Candle Works opened. Known locally as the Candle House.

```
Fatal Accident in Shale Mine

15 Dec 1886 - Broxburn.
   Michael   Kelly,   56,   trimmer.
Crushed by wagons at screens.
```

By 1887 the shale miners had formed a union and wanted company recognition for this. They also wanted an increase of 2d per ton of shale mined. Broxburn Oil Company refused. There was a five-month strike, during which forty three families were evicted from company houses. The strike ended with Broxburn Oil Company recognising the union, and giving the miners their 2d per ton extra.

```
Fatal Accident in Shale Mine

7 Jul 1887 - Broxburn.
   Andrew   Beith,   contractor,   and
William Wilson, oversman.
   They went into a place which had
not been examined, and gas fired at
their naked lights.
```

```
Fatal Accident in Shale Mine

8 Oct 1888 - Broxburn.
   Richard  Sneddon,  57,  oversman.
Struck with tubs on incline.
```

Fatal Accident in Shale Mine

15 Dec 1888 - Broxburn. [Stewartfield No 1 Mine]

Robert Anderson, 15, drawer. Went into fenced-off upset, where there was gas, with naked light.

On Saturday, at half-past one o'clock, an explosion of firedamp took place at No 1 Stewartfield mine, and Robert Anderson, aged 16[sic], was killed, and another man named Thomas Joyce badly injured. Anderson, who had heard the report, went to explore the scene of the explosion, and was killed by afterdamp.

(The Scotsman, 17 December 1888)

Fatal Accident in Shale Mine

29 Jul 1889 - Hayscraig.
William Bell, 40, miner. Shale fell on him from the roof.

1890 - Hut Shale Pit closed.

1891 - The new Henderson retort replaced the 1873 version.

```
          Near-Fatal Accident in Shale Mine

    24 Mar 1891 - Broxburn.
      James Bonnar, 22, drawer. Crushed
    by a tub.
      This accident happened on a cut
    chain brae in the Broxburn Oil Shale
    mine, and was caused by the bottomer
    signalling for a loaded hutch to be
    sent down the brae before he had
    received the signal from the last
    bench further up the brae that had
    made use of the chain, that the
    empty hutch was detached, and the
    cut fixed; the consequence was that
    the full hutch ran amain with only a
    portion of the chain attached, and
    striking the empty hutch attached to
    the chain at the foot of the brae,
    caused the empty hutch at the bench
    higher up to move forward suddenly
    and injure the drawer there. the
    bottomer contravened Special Rule 81:
    he was charged before the sheriff,
    pled guilty, and was admonished.
```

```
         Fatal Accident in Shale Mine

    21 Nov 1891 - Broxburn.
      John Stewart, 50, miner. Shale
    fell on him from the roof.
```

1892 - Mines and oil works opened by Broxburn Oil Company at Roman Camp.

1894 - Part of Albyn Shale Mine (Broxburn Shale section) closed.

1895 - The Sulphuric Acid Plant opened at Broxburn Refinery.

Fatal Accident in Shale Mine

14 Jan 1897 - Broxburn.
Joseph Ellis, 30, miner.
Deceased was mending the holing of shale, in which two shots had just been fired, when it fell over on him; the shale was not spragged.

Fatal Accident in Shale Mine

7 May 1897 - Broxburn.
James McNee, 50, miner.
Deceased was driving a 12-feet place in the "Curly" seam, about 6 feet above the stooped waste of the "Broxburn" seam. The fireman detected a break in the roof which he considered to be dangerous, and requested him to set a prop to it. Deceased demurred, but as the fireman insisted, he agreed. While he was setting the prop, the roof gave way, killing him instantly.

Fatal Accident in Shale Mine

25 Jan 1898 - Broxburn.
David Ferguson, 44, chain runner.
While ascending an In-going-eye mine on a loaded rake of tubs, deceased's body came in contact with some "crowns" on the roof.

Fatal Accident in Shale Mine

12 Jun 1899 - Broxburn.
Daniel Murray, 13, trapper.
Deceased was sitting leaning against a prop on the inside of a sharp curve where a level road joined an engine plane at right angles. A set of nine loaded tubs was being lifted by the haulage rope, and, owing to the extra strain due to one of the rear tubs having become derailed, several of the leading tubs tilted over towards the inside of the curve. One of them struck deceased, crushing his head against the prop, and killing him instantly.

Fatal Accident in Shale Mine

3 Oct 1900 - Broxburn.
William McLauchlan, 21, miner.
Deceased was working off some loose shale at the face of a five yard room in a 6ft seam, when about 14 cwts of the shale came away unexpectedly from some lypes and struck him on the head, causing injuries which terminated fatally immediately afterwards.

Retorting and Refining at Broxburn

This Report appeared in the West Lothian Courier on Friday, 21 November, 1902

Different works have different kinds of retorts, though all agree more or less in their fundamental principle. The retorts in use in the Broxburn works are Henderson's improved patent, by the introduction of which a larger yield of sulphate of ammonia is obtained, without the quality or the quantity of the crude oil suffering in any way.

Up till about eleven years ago the retort exclusively used by the company was the retort patented in 1873 by Mr Henderson, which in its day enjoyed a full share of patronage. The principal features of this retort were the utilisation of the carbon in the spent shale and the gas resulting from distillation as fuel, and also the adoption of a downward distillation. The new retorts, of which there are five hundred in use at the works, are vertical, and are about 28 feet long. The upper portion is about 11 feet long, and is made of cast iron, while the lower portion is built of brick. The shale having been conveyed to the top of the retorts, is emptied into malleable-iron hoppers attached to them. These hoppers are capable of holding eighteen hours' supply of raw shale, and mechanically keep the retorts supplied, being also so constructed as to dispense with night and relieve Sunday work. In the upper portion of the retorts the oil is first distilled off the shale at a temperature of 900 degrees Fahr., while the spent shale passes downward into the brick portion of the retort, being kept in constant motion by a roller at the bottom, which delivers it into a hopper below at the required rate. This portion is kept at a temperature of about 1300 degs. Fahr., the higher temperature largely increasing the yield of ammonia and permanent gases from the carbon of the shale.

The hot products of distillation are drawn from the top of the retorts in a form of oil and watery vapours and permanent gases, passed through a long range of cooling pipes, and the condensed liquors, on being separated, are called crude oil and ammonia water.

The incondensable or permanent gas is used as a fuel, while the spent shale, being of no further use, is stored in a large bing. The condensed fluids flow into a small tank separator, where they are separated by simply taking advantage of their specific gravities. The ammonia water flows off through a pipe fitted close to the bottom of the separator, while the crude oil flows off in another direction through a pipe near the top.

The crude oil, which is of a greenish hue, is now pumped to the refinery, which, as already stated, is about a quarter of a mile distant. The ammonia water is sent to the ammonia house, and there converted into sulphate of ammonia. The method of manufacture of sulphate of ammonia at Broxburn works is a process which has been developed by Mr Henderson, the works manager. The water is first introduced into Mr Henderson's patent column still, which is provided with a series of specially constructed trays. Entering at the top, the water passes from tray to tray until it is discharged at the bottom of the still, where steam is introduced by a pipe from a boiler at a pressure of about fifty pounds per square inch. The steam forces itself through the ammonia water in the lower trays and escapes at the top of the still, carrying with it ammonia gas into a separator containing sulphuric acid. Bubbling through holes in a series of lead pipes at the bottom of the saturator, the ammonia gas combines with the sulphuric acid and forms sulphate of ammonia.

The steam, carbonic acid gas, and sulphuretted hydrogen are conveyed back to the retorts, and the steam used again.

As the sulphate of ammonia is formed, it is "fished" out with perforated ladles and thrown into a large bunker, where it is allowed to remain for twelve hours to drain. It becomes solidified as it dries, and after passing through the storehouse, it is put through the mill and bagged for the markets, where there is a good demand for it as a fertiliser.

The Refining Process

One is glad to leave the rank-smelling retorts and pass to the cleaner refinery. What with climbing upstairs to the top of the retorts, walking along planks, which, though thoroughly safe, make one shiver at the consequences if he were to fall down, the visitor has by this time become somewhat tired, though his interest is hardly likely to be diminished. His hands are of the same colour as the sides of the retort, and he feels that his complexion is such that his own mother would not know him.

The refinery is like a well-tidied kitchen. The crude oil is received at the refinery in a large tank capable of containing about 50,000 gallons. The object of the refinery is to clear the oil of all impurities, and this is accomplished by repeated distillations, during which it is fractionated into the different gravities required; and by treatment with oil of vitriol and with caustic soda.

The distillations are according to Henderson's patent continuous system. The first distillation gives green naphtha, green oil, and still roke, which is a valuable smokeless fuel.

The green naphtha is treated with oil of vitriol and with caustic soda, and after another steam distillation, is ready for the market as shale spirit or naphtha with a specific gravity averaging 730 to 740.

The green oil is treated with oil of vitriol and caustic soda, and is then ready for the second distillation. In the treatment, the oil is stirred first with oil of vitriol and allowed to settle, when a black tar falls to the bottom and is run off. The oil is run by gravitation into another vessel, and is mixed with solution of caustic soda. On settling, another black tar separates out. The chemicals are recovered from the tarry matters, and the residue tars are used as fuel by being injected with steam into the still furnaces in the form of fine spray, and burned.

The recovered sulphuric acid is used for making sulphate of ammonia. The oil of vitriol used in the oil refining is manufactured by the company.

In the second distillation, the green oil is fractionated into light oils - the latter containing solid paraffin. The former are treated and distilled several times, and are ready for the market as burning oils. The heavy oils are cooled below the freezing point of water, and the paraffin strained out by passing through filter presses - the two products from the filter presses being now termed blue oil and solid crude paraffin.

The blue oil is treated and distilled, (then) fractionated, (producing) intermediate oils - used for gas making and other purposes - and heavy oils for lubricating machinery.

The crude solid paraffin is now passed through filter presses and then sent to the sweating chambers, where it is subjected to Mr Henderson's patent for purifying paraffin, and is fractionated and refined from 100 degrees to 120 degrees Fahr. melting point, and is then known as semi-refined paraffin. Should it be desired to further refine this paraffin or make it pure wax, the sweating and fractionating process is repeated, and the paraffin is then known as refined paraffin wax, of from 125 to 130 degs. Fahr. melting point.

After the sweating process, the liquid paraffin is mixed with char, settled, and then filtered through paper, when it is cooled into blocks of paraffin wax, and is then ready for making into candles, night lights, wax vests, etc. The soft paraffin, with a melting point of 100 degs. Fahr., is called match paraffin, and is used, instead of brimstone, for dipping the ends of wood matches, while it is also used for burning in mines and other lamps, and numerous other purposes.

Not the least interesting of the many departments is the candle factory, situated at the south-west of the refinery. Here are large stacks of candles of various shapes and sizes. Each of the machines is capable of turning out 300 candles, and the total output amounts to about nineteen or twenty tons of candles per day. The boxes in which the candles are packed are made on the premises by packing case nailing machinery, and almost before one has time to realise what is being done, a neat box is put together. The candles are sent all over the world.

The work also comprises an extensive cooperage, smiths' shop, pattern shop, and joiners' shop and sawmill.

The company have also crude oilworks at the Roman Camp, where there are 240 of Henderson's improved patent retorts in use, and, in the Drumshoreland seams of shale, the yield of crude oil is twenty gallons to the ton of shale, whilst the yield of sulphate of ammonia is 60 to 70 lbs. to the ton. At present the company have on hand a scheme for renewing their retorts at Broxburn, and erecting retorts where the through-put per retort will be twice as large as in those in existence. Already operations are in progress for the erection of one bench.

Our representative is indebted to Mr Norman M. Henderson for permission to see the work, and to Mr D. R. Steuart, the chemist, for so kindly showing him round and explaining the different processes.

Broxburn Oil Company, 1903-1919

1903
Stewartfield No 1 Shale Pit closed.

1905
The remainder (Curly Shale section) of Albyn Shale Mine closed.

Fatal Accident in Shale Mine

27 Mar 1905 - Broxburn.
Michael Brogan, 32, miner.
The holing is in the centre, and bottom shale is taken away first, afterwards the top shale, which generally overhangs and is supported by sprags. Deceased was found under the top shale which came away from a "dry" parting. There were no sprags up to the shale.

Fatal Accident in Shale Mine

22 Aug 1905 - Broxburn.
Robert Practice, 38, miner.
The top shale fell away suddenly on deceased, killing him instantly.

Fatal Accident in Shale Mine

3 Oct 1905 - Broxburn.
Hugh McCord, 37, miner.
Deceased was taking down the top shale when it fell away and fatally crushed him.

Fatal Accident in Shale Mine

4 Oct 1906 - Broxburn.
George Shanks, 32, miner.
Deceased was found 16 feet from his working place injured on the head by being struck with flying shale from a shot. The shot was in the bottom portion of the seam, and was charged with 1lb of gunpowder, and it appeared deceased had been delayed after the fuse was lit and failed to reach a place of safety when the shot exploded. His lamp was found near the shot, which indicated that probably the light had been extinguished by the spitting of the fuse, and he had dropped it and put off some time searching for it.

Fatal Accident in Shale Mine

15 Oct 1907 - Broxburn.
John McVey, 42, miner.
Deceased was pinching off bottom shale, when the roof suddenly collapsed and buried him.

Fatal Accident in Shale Mine

1 Mar 1909 - Broxburn.
Robert Anderson, 36, oversman.
The mine starts from the surface, and the mineral is hauled by an engine and dook rope. A loaded rake had been hauled a short distance up a steep gradient, when a chain bar of one of the tubs broke and 8 tubs came back, and deceased was fatally crushed.

```
     Fatal Accident in Shale Mine

   7 May 1909 - Broxburn.
   Robert Kennedy, 27, roadsman.
      The drawer at the top of a "Cuddie
   brae" allowed his loaded tub to run
   too far, and as the blocks were out
   it ran on to the brae, deceased was
   proceeding down at the time, and on
   hearing the shout to stand clear
   he stepped right in front and was
   carried to the bottom, and fatally
   crushed.
```

```
     Fatal Accident in Oil Works

      20 Dec 1910 - Albyn Oil Works: No
   4 Bench Retorts.
      William   Crawford,   retort   fans
   man, Albyn Hut Cottage, Broxburn,
   was injured when caught between the
   spokes of a fly wheel of an engine
   and the floor.
      Died in the Royal Infirmary the
   same day.
```

1912 - Hut Shale Mine closed.

The shale-oil industry began to experience major problems during World War I because of serious competition from the Anglo-Persian Oil Company Limited and from imported American oil. Neither the Americans nor the Anglo-Persian had to incur the costs of mining shale then extracting the oil, so it was a cheaper product to obtain.

In addition to the lower production costs of rivals, another problem the shale-oil industry had was the fact that the British government had a controlling interest, i.e. financial investment, in the Anglo-Persian Oil Company Ltd. They would therefore look after its interests first.

1914
Stewartfield No 2 Shale Pit closed.
Stewartfield No 3 Shale Pit closed.
Stewartfield No 4 Shale Pit closed.

1915
South Greendykes Shale Mine closed

Fatal Accident in Oil Works

20 Jan 1916 - Broxburn Oil Works. John Cushley, labourer, 169 Back Street, Broxburn, injured in sulphate house when he fell into a grinding mill.
Died at his house on 22 January.

Fatal Accident in Oil Works

24 Oct 1916 - Broxburn Oil Works. John Millar, labourer, 86 Stewartfield Rows, died when he was struck and then fell into a grinding mill in sulphate house.

1918
Excise Duty relief of 6d per gallon on home-produced light hydrocarbon oils was introduced.

The Scottish Oil Agency & Scottish Oils

On 16 May 1918, the Scottish Oil Agency Ltd was formed by William Fraser of Pumpherston Oil Company Ltd. The five surviving oil companies were members:

Broxburn Oil Company
Oakbank Oil Company
Pumpherston Oil Company
James Ross and Company, Philpstoun Oil Works
Young's Paraffin Light and Mineral Oil Company

The Scottish Oil Agency was created as a selling and distribution business for the products of the above companies. Its address was 53 Bothwell Street, Glasgow.

Certificate of Incorporation of The Scottish Oil Agency

On 16 August 1918, Broxburn Oil Company registered a change in its business activities with a 'Memorandum of Association with respect to its Objects' - a legal document stating the new objectives of the company within the Scottish Oil Agency.

Certificate of Registration .. confirming Alteration of Objects

In 1919, the price of oil fell because of increased American production and exports.

On 6 June, 1919, the following appeared in the *West Lothian Courier*:

'Broxburn Oil Company have also intimated the closing down of one of their mines - their oldest mine in operation, known as Haycraigs - where 90 miners and oncostmen are losing their places. These are, however, to be accommodated as far as possible in the Company's Dunnet Mine.'

1919
Hayscraigs Shale Mine closed.
Hayscraigs Shale Pit (Pyothall No 5) closed.

```
Fatal Accident in Shale Mine

21 Aug 1919 - Broxburn.
Alexander Munro, shale miner of
East Main Street, Broxburn, died on
21 Aug 1919 at Royal Infirmary from
injuries sustained on 20 Aug 1919
when a large stone fell on him.
```

3 Sept 1919 - Scottish Oils Ltd was incorporated. It was created by the Anglo-Persian Oil Company Ltd as a holding company.

The following five companies came under the management of Scottish Oils, and, as the holding company, Scottish Oils held (owned) their shares.

Broxburn Oil Co Ltd
Oakbank Oil Co Ltd
Pumpherston Oil Co Ltd
James Ross and Co, Philpstoun Oil Works Ltd
Young's Paraffin Light and Mineral Oil Co Ltd

On 2 October 1919, Scottish Oils was admitted as a member of the Scottish Oil Agency.

No. 10612

Certificate of Incorporation.

I hereby Certify, That "SCOTTISH OILS, LIMITED" is this day incorporated under the Companies Acts, 1908 to 1917, and that this Company is **Limited.**

GIVEN under my hand at Edinburgh, this ___Third___ day of ___September___ One Thousand Nine Hundred and Nineteen.

Wm Vickers

for Registrar of Joint-Stock Companies.

Certificate of Incorporation of Scottish Oils

No. 10612.

Certificate under s. 87 (2) of the Companies (Consolidation) Act, 1908 (8 Edw. VII. c. 69), that a Company is entitled to commence business.

I hereby Certify, That

"SCOTTISH OILS, LIMITED"

which was Incorporated under the Companies Acts, 1908 to 1917, on the Third day of September 1919, and which has this day filed a statutory declaration in the prescribed form that the conditions of s. 87—1 (a) and (b) of the Companies (Consolidation) Act, 1908, have been complied with, is entitled to commence business.

GIVEN under my hand at Edinburgh, this Second day of October One Thousand Nine Hundred and **Nineteen**.

Wm Vickers
for Registrar of Joint-Stock Companies.

Certificate, that Scottish Oils is entitled to commence business

Scottish Oils Limited

Incorporated 3 September, 1919

Registered Office - 7 West Nile Street, Glasgow
Nominal Capital - £1,000,000 divided into 1,000,000 shares of £1 each
Company Subscribers -
Anglo-Persian Oil Company Limited, 23 Great Winchester Street, London EC2 (991,998 shares)
Sir Charles Greenway, Merchant, [Chairman & Managing Director of the Anglo-Persian Oil Co] 23 Great Winchester Street, London EC2 (1,000 shares)
Robert Irving Watson, Merchant, 23 Great Winchester Street, London EC2 (1,000 shares)
Sir Frederick William Black, Merchant, 23 Great Winchester Street, London EC2 (1,000 shares)
John Douglas Stewart, Shipbroker, 23 Great Winchester Street, London EC2 (1,000 shares)
John Buck Lloyd, Merchant, 23 Great Winchester Street, London EC2 (1,000 shares)
William Fraser, Merchant, 135 Buchanan Street, Glasgow (1,000 shares)
Hubert Edward Nichols, Merchant, 23 Great Winchester Street, London EC2 (1,000 shares)
Duncan Garrow, Merchant, 23 Great Winchester Street, London EC2 (1,000 shares)
Robert Alexander Murray, Chartered Accountant, 175 West George Street, Glasgow (1 share)
Henry Rowland Cooke, Accountant, Maybury Lodge, Reigate (1 share)
[The headquarters of Scottish Oils Ltd was Middleton Hall, Uphall]

Broxburn Oil Company, 1920-1925

Although now under the management company Scottish Oils, and also a member of the Scottish Oil Agency, Broxburn Oil Company still ran its own affairs on a daily basis - up until World War I, when mines came under government control. After the war, the mines were returned to private ownership. There was an increase in costs (there was high inflation, and the price of coal had gone up), and in 1920 there was a strike by coal miners during which some shale mines were closed.

```
Fatal Accident in Shale Mine

25 Feb 1920 - Dunnet.
   William Crawford, braker, 78
Stewartfield Rows, died when he was
struck by a haulage rope.
```

The situation had changed. In 1920, shale miners had been paid £6 for a six-day week. Now drastic cuts were made, and in 1921 their wages were halved to £3 per week. A strike followed, and Broxburn Refinery, Albyn Oil Works, Stewartfield No 1 Shale Mine, and Dunnet Shale Mine closed. Soup kitchens were set up, and gave families one meal per day. The men went back to work after six long months - with no increase forthcoming.

```
Fatal Accident in Shale Mine

23 Feb 1921 - Dunnet.
   James Potter, miner, 2 Shrine
Place, Broxburn, died when a quantity
of shale fell on him.
```

1921
The Excise Duty relief of 6d per gallon on home-produced light hydrocarbon oils (1918) was now withdrawn.
(Withdrawn from 1921 to 1927)

On 20th January 1922, the following appeared in the *West Lothian Courier*:

'The announcement that a definite start had been made this week towards the re-opening of the Broxburn Oil Works was everywhere hailed with satisfaction. It will take some time to get the retorts put in order, and it is unlikely that even one bench will be working much earlier than the end of February, but the prospect of what may be has brought hope. The community will have cause to feel thankful if April should bring a general resumption. The miners are deeply concerned as to whether the working of Dunnet and Carledub's Mines is to be resumed.'

```
Fatal Accident in Oil Works

12 Apr 1923 - Broxburn Oil Works.
  Robert Miller, foreman oilworker,
Bridge Place, Broxburn, injured when
he slipped and fell astride a pipe.
He died on 19 April 1923 at the
Royal Infirmary, Edinburgh, from his
injuries.
```

1923 - Stewartfield No 1 Shale Mine closed.

```
Fatal Accident in Oil Works

20 Feb 1924 - Broxburn Oil Works.
  Richard O'Hare, labourer, 22
Stewartfield Rows, Broxburn, injured
when he fell from a wooden platform
to the ground.
  Died 3 July 1925 at Longmore
Hospital for Incurables, Edinburgh.
```

```
Fatal Accident in Shale Mine

3 Sep 1924 - Dunnet.
David Given, miner's drawer, 79
Stewartfield Rows, Broxburn, was
struck on the head when a piece of
shale came away from the roof, and
died in a few minutes.
```

```
Fatal Accident in Shale Mine

25 Sep 1924 - Dunnet.
George Sibbald, mine roadsman, 6
Shrine place, Broxburn, died when he
was struck by a piece of shale from
a shot on 24 Sept 1924.
```

In 1925 there was a six-week strike to fight against another proposed wage reduction of 10%. The shale-oil companies were losing money. Scottish Oils was making a profit. The oil workers trade union maintained that the five shale-oil companies and Scottish Oils were one and the same company (see 3 Sept 1919). Scottish Oils disagreed. *A Court of Inquiry into the dispute found in favour of Scottish Oils. They decided that Scottish Oils Ltd and the shale-oil companies were separate businesses. Therefore, Scottish Oils had no reason to subsidise shale oil losses from their other commercial enterprises.* [Presumably from substances such as ammonium nitrate, coke, wax products (candles, waterproofing, matches, petroleum jelly, etc) sulphuric acid, et al.]

The Broxburn Shale Mines and Oil Works closed at Martinmas - St Martin's Day, 11 November - 1925.

They Never Opened Again.

Broxburn Oil Works

Opposite: two views of Broxburn Oil Works around 1900.

Above: Albyn Oil Works c. 1900.

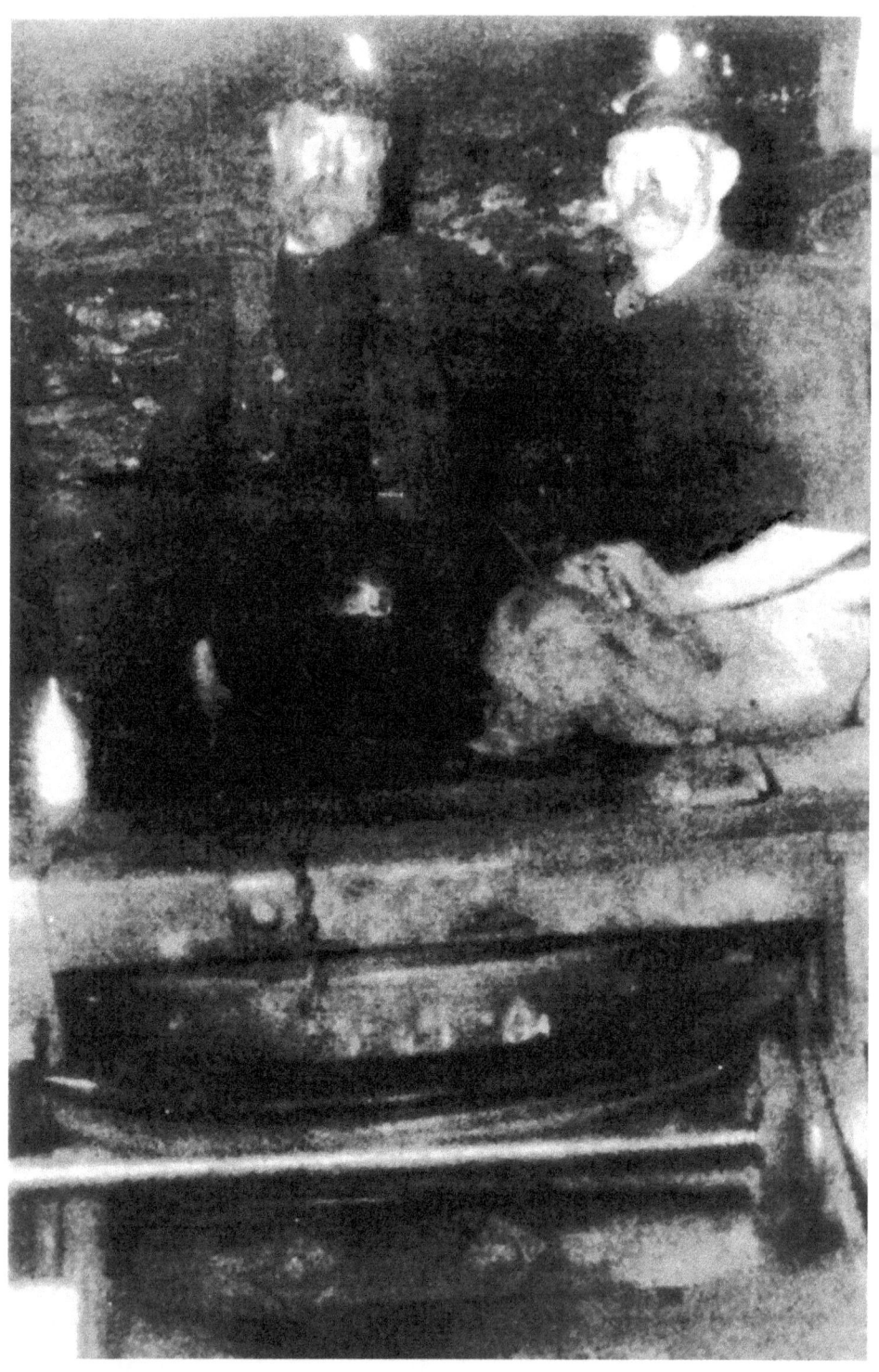

Miners in Hayscraigs Shale Mine, c. 1910. Despite the danger, the men are using oil-flame lights, because the Davy lamp emitted little light.

Man at the top shot boring on an incline, c. 1900.

Opposite, top: Railway within Broxburn Oil Works, c. 1910.

Opposite: Broxburn Oil Company Locomotive, c.1920.

Above: The Iron Bridge over west Main street from the canal bridge, c. 1905.

The mineral railway ran to shale mines and oil works at Broxburn, Holmes, Roman Camp, Uphall, Pumpherston, and Winchburgh, so all were linked by this local rail network. It connected with the Edinburgh-Glasgow main line near Winchburgh, and with the Bathgate line at Drumshoreland.

It ran across the Iron Bridge from Broxburn Oil Works to Drumshoreland, where coal was collected and taken to the oil works, or via a lye — a small branch line — to Holygate coal and goods yard — the white hut just visible in the middle distance on the left.

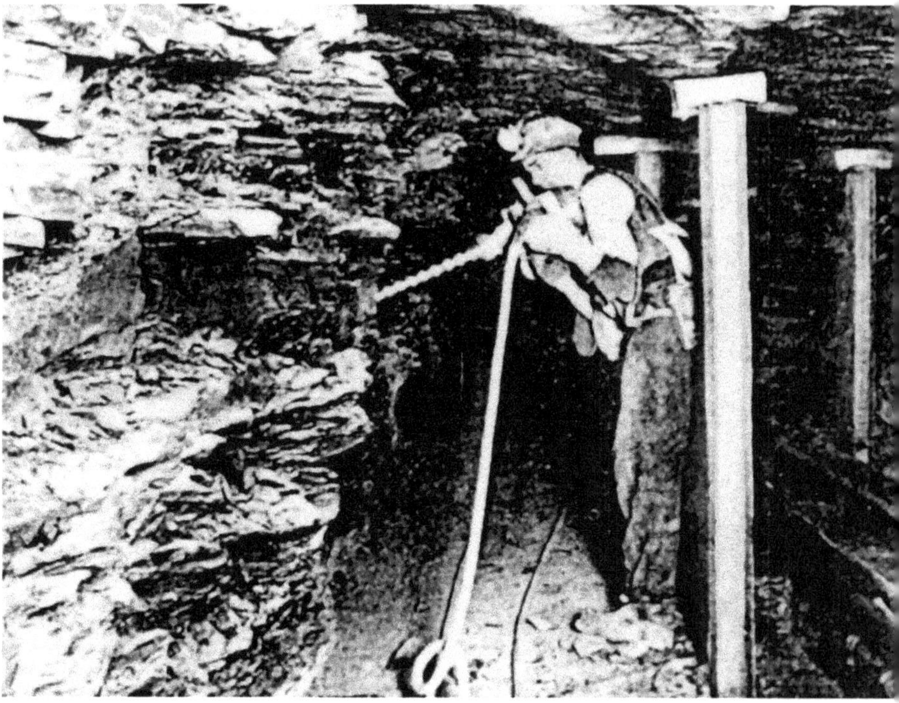

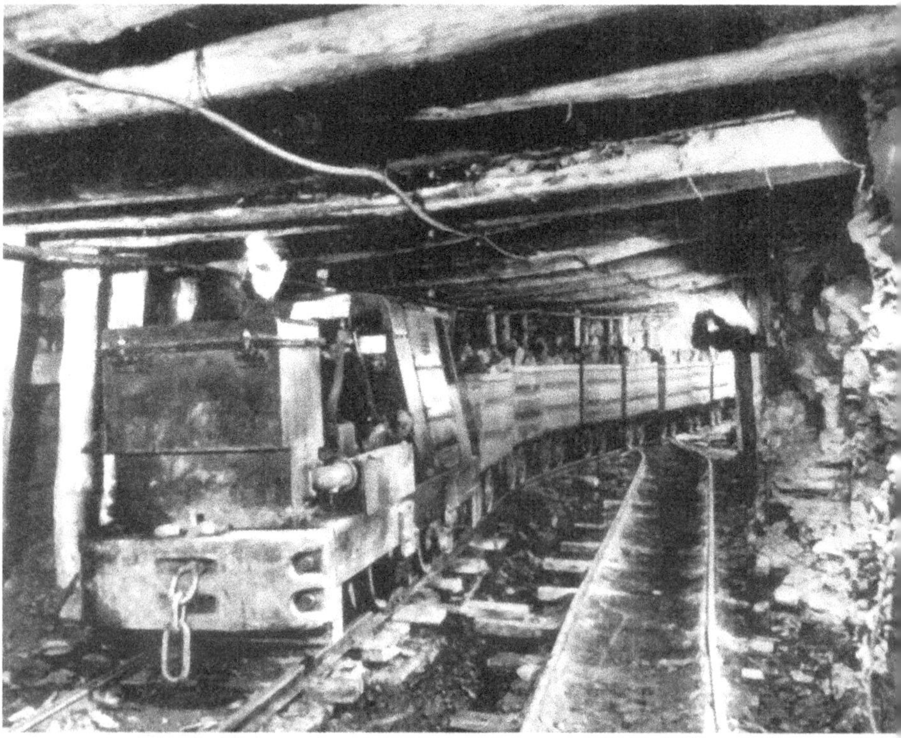

Opposite, top: Boring a shot-hole in which to place the charge with compressed air drill, c.1910.

Opposite, below: Diesel locomotive pulling hutches loaded with shale, c.1915.

Above: Broxburn Oil Works, c. 1910. The corner of Albyn shale bing is visible in the right foreground.

View of Broxburn Oil Works from Broxburn Gas Works, c.1910.

Broxburn Candle House, in the 1930s. This stood on the north side of the Union Canal at Greendykes. The candles were made from paraffin wax, a shale oil by-product.

Advertisement for Broxburn Paraffin Candles.

Greendykes Road.

Above: c. 1908. The chimneys, left, of Broxburn Oil Works; right, Broxburn Gas Works.

Opposite, top: 1916. The houses were built by Broxburn Oil Company between 1878 and 1884.
By the way the two women are dressed, it is a Sunday morning. They may well be walking back from church to the detached villas of the oil company management, further up Greendykes Road.

Opposite, below: c. 1925. Although these houses were called miners' rows, all oil company workers — not just the shale miners — lived in them.

Above: Broxburn Oil Company's houses at Greendykes, 1960. They were demolished in the late 1960s.

Opposite, top: Stewartfield Rows, c.1935. Built by Broxburn Oil Company in 1883-84 on the land of Stewartfield Farm, they were demolished in 1966.

Opposite, below: The old Oil Company houses, showing the proximity to the Stewartfield tip, c. 1960.

Opposite: Houses in the soon-to-be-demolished Stewartfield Rows in the 1960s.

Above: A view of what remained of the Broxburn shale bings in 1982. Left to right are Greendykes tip, Albyn tip, Stewartfield tip.

Albyn tip, now gone, beside Bridge No 29 in 1996.

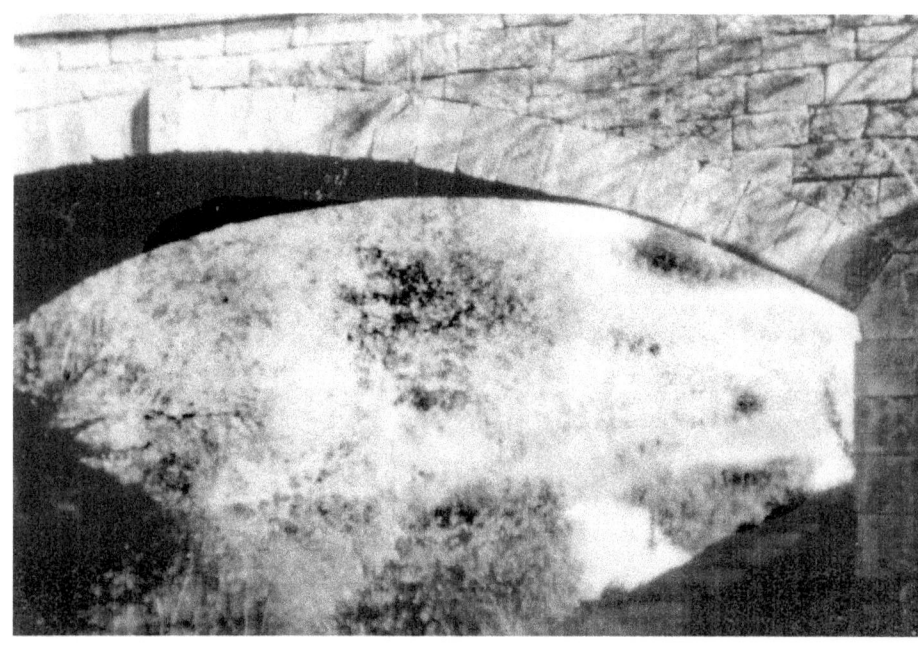

Bridge No 28, where the Broxburn shale industry began, when Robert Bell built his first shale oil works in 1862.

An aerial view of the Broxburn shale bings, in 1947. The Union Canal winds its way between the Stewartfield and Albyn shale bings. The former miners' rows of Greendykes and Stewartfield are still inhabited, although the oil works and shale mines in which their original occupants worked have been gone for some twenty years.

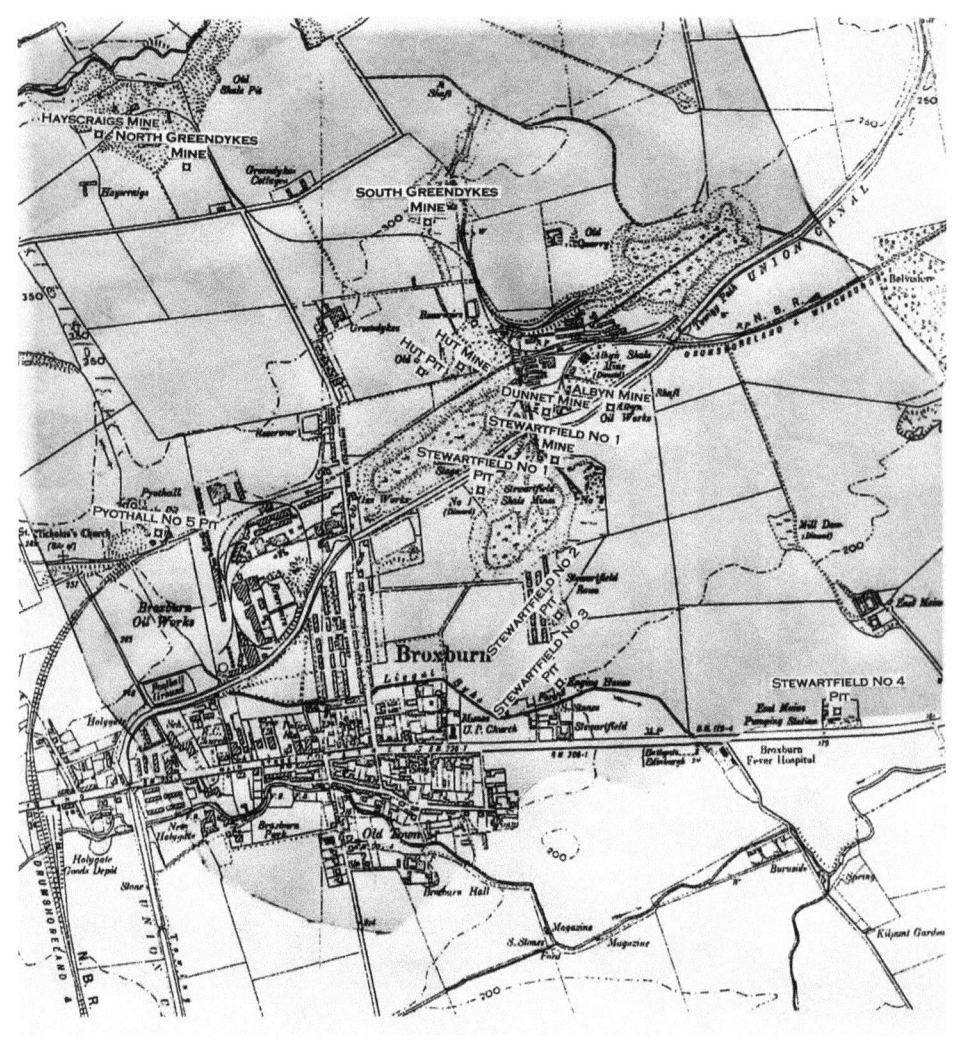

The approximate sites of the entrances to the Broxburn shale mines and pits. The shaded area shows the extent of the known underground shale mining.

A Question of Profits

The five oil companies (Broxburn Oil Co - Oakbank Oil Co - Pumpherston Oil Co – James Ross & Co - Young's Paraffin Light & Mineral Oil Co) were in business as individual companies.

They were members of the Scottish Oil Agency, which was itself a company. Despite the name 'Agency' the Scottish Oil Agency was incorporated as a company, just like the others.

The five individual oil companies 'joined' this other company. A modern example would be someone in the public eye employing an agent to publicise and market them, i.e. get them as much money as possible, the agent being paid a percentage.

So, the companies were clients of the Agency, just as the modern celebrity is a client of his/her agent. The Scottish Oil Agency would have been paid a percentage of the sales when they marketed and sold the oil products of the five companies.

In the 1920s, just as now, oil was a very important product. There were a lot less cars than nowadays, but, unlike now, there were mills, factories, engineering works, iron foundries, steel works, and a multitude of manufacturing industries all over Britain – and all required oil in order to operate their machinery.

Again, unlike now, there were farms all over the country growing crops (many of the former farm lands now have housing estates built on them), and ammonium nitrate was in demand as a fertilizer.

Also, every home in Britain used paraffin and candles in their homes. Yet the oil companies were losing money.

Scottish Oils was incorporated by the Anglo-Persian Oil Company to hold the shares of the five oil companies which now came under its management.

Scottish Oils was set up by the Anglo-Persian Oil Company as a holding company. The Anglo-Persian was the parent company of Scottish Oils.

The terms Parent Company and Holding Company seem

to have the same meaning, but a Parent Company would be one which has its own assets, say a steel manufacturing business, then it buys up the shares of another company, thereby becoming its Parent Company, while at the same time carrying on its own steel production business. The other company's business carries on as well, but it is now under the control of the Parent Company – it has become another asset of the Parent Company.

A Holding Company has no assets such as a manufacturing business, because it is set up solely to hold shares of other companies. Once it has acquired the shares of the other companies, these other companies become its assets.

Scottish Oils became the holding company for the shares of Broxburn Oil Company, Oakbank Oil Company, Pumpherston Oil Company, James Ross & Company, and Young's Paraffin Light & Mineral Oil Company.

If you own the shares of a company, you own the company.

The five oil companies were now under the management of Scottish Oils. This meant:

There was now one board of directors managing the five oil companies (and Scottish Oils) - the board of directors of Scottish Oils.

Scottish Oils joined the Scottish Oil Agency.

The oil companies were losing money. Scottish Oils was making a profit. The oil union believed that the profits being made on some products should have been used to offset the losses being made on others. Scottish Oils disagreed. The dispute was decided by a Court of Inquiry.

A Court of Inquiry was not what is usually understood by the definition of the word court, i.e. a court of law. It would have been a panel appointed to decide on the matter in question. Who appointed the panel, and who sat on it, is undetermined.

The Court of Inquiry decided that the oil companies and Scottish Oils were all individual companies, and consequently, Scottish Oils did not have to compensate from their profits for the losses incurred by the individual oil companies.

But it is more complicated than the fact that Scottish Oils was a company in its own right. As stated previously, Scottish

Oils was a holding company, and was formed for the purpose of controlling the other companies.

If Scottish Oils owned the shares of Broxburn Oil Company *et al* it owned Broxburn Oil Company. Scottish Oils as a company did not own shale mines or oil works *per se*, but, through its board of directors, it owned companies which did own shale mines and oil works. The oil companies became subsidiary companies to Scottish Oils, which was created by the Anglo-Persian Oil Company solely for this reason.

[Government-controlled] Anglo-Persian Oil Company
|
Scottish Oils
|
Broxburn Oil Company
Oakbank Oil Company
Pumpherston Oil Company
James Ross and Company
Young's Paraffin Light and Mineral Oil Company

The Anglo-Persian Oil Company was the parent company of Scottish Oils, and Scottish Oils was the holding company of Broxburn Oil Company.

So, it wasn't just as straightforward as the Court of Inquiry made it out to be - that they were all individual companies, and that was the end of the matter.

Scottish Oils didn't produce shale or oil, Broxburn Oil Company did that. Broxburn Oil Company was the company mining the shale, processing it through their oil works, then through their refinery, finally producing the various oil products which were marketed and sold by the Scottish Oil Agency. But Broxburn Oil Company was no longer in control of its own oil business - which it had been before 3 Sept 1919. It was now run by its management company, Scottish Oils.

But, if Scottish Oils was making a profit, surely Broxburn Oil Company should have been making a profit. After all, that's where Scottish Oils' profits came from.

On the surface, the decision to close Broxburn Oil Works and shale mines would seem to have been taken by the board of Scottish Oils, but Scottish Oils being a subsidiary of the Anglo-Persian Oil Company, the decision would have been made for them in a boardroom at 23 Great Winchester Street, City of London. However, Anglo-Persian was government controlled, so the decision to shut down and demolish Broxburn Oil Works originated from an office in Whitehall. The bottom line is that it was the decision of politicians in London.

The Anglo-Persian Oil Company and British Petroleum

In May 1901, 51-year-old English-born Australian businessman William Knox D'Arcy was granted a concession by the Shah of Persia to explore for oil. By 1905 D'Arcy was running low on funds and needed a cash injection. The Rothschilds, on behalf of the French Government, were prepared to sign a contract. An oil committee including representatives of the British company Burmah Oil and a young Winston Churchill are said to have agreed to give D'Arcy financial support of £1,000,000. In return, when oil was found he agreed to form a company named Anglo-Persian Oil, with the oil concessions going to the British through the Burmah Oil Company. In May 1908 a large Persian oil-field was struck.

On 14 April 1909, the Anglo-Persian Oil Co was incorporated, William Knox D'Arcy being appointed as a director. The Abadan Oil Refinery was built, and production began in 1913.

By 1914, Anglo-Persian was in financial trouble, because, strangely, although it had plenty of oil, it did not have enough customers to sell it to. Cars were few, so they were not a major market, and European and early US oil companies, having been up and running for many years, had cornered the market for industrial oils - the Persian oil could not be sold for heating homes because it had a sulphurous smell.

That year, the First Lord of the Admiralty, Winston Churchill, convinced Parliament that the British government should become the major shareholder of Anglo-Persian so that Britain could be sure of obtaining oil for its new oil-fired warships.

(Burmah Oil Co, who had an oil lease from the Burmese Government, had refineries at Rangoon and oil works in Burma and India. They held almost all the ordinary shares — just under a million — in Anglo-Persian. Their headquarters, like that of Anglo-Persian, was in Great Winchester Street, City of London. William Knox D'Arcy was now also a director of Burmah Oil.)

The government injected £2,000,000 into the company, buying a 51% shareholding. The Anglo-Persian cash problem was solved. The British government had gained controlling interest, and were now the real power behind the company. Two months later, World War I began.

The origins of British Petroleum are rather surprising. British Petroleum (registered 21 Nov 1906) was a subsidiary of the German company Europaische Petrolem Union who were controlled by Deutsche Bank. The name 'British Petroleum' was the name under which Europaische Petroleum Union operated in Britain. It was a name created to promote and sell its oil products here. During World War I, the government seized the British assets of the German company - 150 depots, 535 rail tank-wagons, 1,102 road vehicles, 4 barges, and 650 horses - and in 1917 they sold these (including the German company's former British marketing name, British Petroleum) to the Anglo-Persian Oil Company, who now had instant access to an established distribution network throughout Britain. The Anglo-Persian Oil Co then operated in Britain under the tradename British Petroleum Oil Co - just as the Europaische Petroleum Union had done previously - using the trademark BP.

(In 1921, there were only 69 BP petrol pumps in Britain. In 1925, there were 6,000.

In 1935, Persia changed its name to Iran, so the Anglo-Persian did likewise, becoming the Anglo-Iranian Oil Co - their brand name remaining BP.

In 1951, a bill was passed in Iran nationalizing the oil industry, so Anglo-Iranian had to leave. Oil production ceased in Iran. The Anglo-Iranian began oil exploration in countries such as Iraq, Kuwait, and Libya.

National oil production resumed in Iran in 1954. That same year, the company dropped the name Anglo-Iranian Oil Company and adopted the name British Petroleum, the name they had used to sell their products in Britain. Their UK marketing name had now become their company name.)

Official Closure, 1927

In 1926, the General Strike took place in support of the coal miners - they had been threatened with a reduction in wages coupled with an increase in working hours. The General Strike only lasted from 3 May to 12 May. The coal miners themselves stayed on strike. However, by the end of November, through economical necessity, most were back at work.

On 6th May 1927, the following report appeared in the *West Lothian Courier*:

```
        Broxburn's Death-Blow
     Works and Mines Not To Re-Open
   'Consternation reigned in Broxburn
yesterday, when it became known that the
following notice had been posted at the
Works and Mines:-
   "The Broxburn Oil Coy. (Ltd.) regret
to intimate that Dunnet Mine, Albyn Crude
Oil Works, and Broxburn Refinery will not
be re-opened; and notice is hereby given
that all contracts of service with the
Company at these places will terminate on
Saturday, 14th May, 1927. The employment
of any workers who may be engaged in
dismantling or otherwise, will be from
day to day only."
   By Order, R.W.Meikle, Secretary,
                          5th May, 1927.
```

Previous to the four weeks' strike, which took place at the end of 1925 - to resist a reduction of 10 per cent in rates and wages - about 1200 men were employed at the above works and mine. Since the strike the works and mine have remained closed, only care and maintenance men being employed. A small proportion of the unemployed have succeeded in getting work elsewhere or have emigrated, but a very large number of men would receive the above intimation with feelings of dismay.

Many are elderly men who have spent all their working days in the employment of Broxburn Oil Company, and are now looked upon as quite unfit for other work.

Another unfortunate result of this abandonment will be the effect it will have upon the public rates of the parish of Uphall. The oil companies pay 50 per cent of the total amount collected by the Parish Council, and the demolition of Broxburn works and the abandonment of Dunnet mine will mean a very large reduction in the valuation of the parish, with consequent increased burdens upon those ratepayers remaining. The debt upon the Parish Council on account of payments to able-bodied unemployed already runs into a very large sum of money.'

1927
Dunnet Shale Mine closed (officially).
Actually closed as of 11 Nov 1925.

On 13th May 1927, the following appeared in the *West Lothian Courier*:

```
The refinery at Pumpherston Oilworks is
being carried on, the crude oil from Roman
Camp Works (Broxburn Oil Company) being
dealt with there.
```

A New Phase

A large number of unemployed workmen - said to be about 200 - have received a letter from Broxburn Oil Company in the following terms:- "You will require to remove from the house at present occupied by you, and belonging to Broxburn Oil Company (Ltd.), and that on or before the expiry of one week from this date. This notice is given in consequence of your right to occupy said house having terminated by your ceasing to be employed by said Company."

On inquiry it has been ascertained that it is only those workmen who are in arrears with their rent (and many have paid no rent since November, 1925) who have received the above notice. As far as can be learned, there is no disposition on the part of the unemployed workmen to obey the notice given them.'

Broxburn Oil Works was dismantled - the sulphuric acid plant, candle house, and rail-wagon repairs, remained in business. No single reason accounts for the closure. The cumulative causes are given as:

1: When wartime control of oil came to an end, prices were increased, and so were wages.

2: There was too much oil being produced worldwide (United States, Russia, Persia, Romania, Dutch East Indies (Indonesia), South America) - i.e. cheap oil was now available from abroad.

3: As a result of (1) and (2) the shale oil companies were now making heavy losses.

4: In 1925, the shale oil companies lost their contract to supply fuel for the Royal Navy, which they had done for the past twenty years.

1 and 2 seem to contradict each other. 1 says Companies charged a higher price for their oil when wartime controls were removed. 2 says there was 'too much' oil. Surely the price of oil would have fallen as a result of this, not risen?

If 1 and 2 are at odds with each other, the conclusion that 3 is the result just doesn't add up.

Was it not more likely that the shale oil companies were making losses because their overheads were high, i.e. it cost them more to get their oil - they had to dig the shale out of the ground, then process it in order to extract the oil? Foreign companies took their oil straight out of the ground.

If the world price of oil had fallen (rather than risen) the cost to produce the oil may have been higher than the selling price of the oil.

But another factor was involved. When the shale oil industry was experiencing financial difficulties, instead of assistance, precisely the opposite happened. In 1921, the British government took the decision to remove the Excise Duty relief of 6d per gallon on home-produced oil - i.e. at a time when the shale-oil companies were losing money, the government demanded an extra 6d tax on every gallon of oil produced. The imposition of this extra tax must have been a crushing blow.

Then, on top of the aforementioned, in 1925 circumstance 4 occurred - the shale oil companies (Scottish Oils) lost their twenty-year old Royal Navy contract. The description 'lost their contract' makes it sound as if this event did not involve anyone else but the shale oil companies, but the companies did not actually 'lose' their oil contract, it would have been taken from them by the British government and awarded to another company.

In 1914, Charles Greenway, Managing Director of the Anglo-Persian Oil Company, signed an agreement with the First Lord of the Admiralty, Winston Churchill, to supply fuel oil to the Royal Navy. So, although a contract already existed whereby oil was supplied to the Royal Navy by its subsidiary company (Scottish Oils), the parent company (Anglo-Persian) signed a separate contract to also supply fuel oil. Of course, this oil would be foreign oil, not local, and was probably cheaper. When the shale oil contract was up for renewal, was the whole contract then awarded to Anglo-Persian ?

Probably, after all it was a *government* company. The loss of their naval contract was another major blow to the shale-oil companies - for some, it was fatal.

It is perhaps surprising that any of the oil companies managed to stay in business, but, somehow, some did manage to survive - for a time. Broxburn Oil Company was one - but not in Broxburn.

Although the Broxburn shale industry had now ceased to exist, oddly, the Broxburn Oil Company still ran the Roman

Camp Oil Works (crude-oil works) on a day-to-day basis (transferred to Scottish Oils, 1919), and worked shale mines there. The crude oil would have been refined at Pumpherston Refinery.

Which begs the question: Why not keep Broxburn Oil Works (the largest in the country) open, and close the Roman Camp Oil Works ? A similar question could also apply to the shale mines, for that matter.

However, things were not as they might have appeared to be. Broxburn Oil Company was not in control of its own destiny.

Decisions concerning the company would have been taken by people with no direct connection to the oil company, or to Broxburn.

Conclusion

Broxburn Oil Company's fate appears to have been effectively sealed in September 1919 when it came under the ownership of Scottish Oils who were owned by the Anglo-Persian Oil Company, foreign oil producer and major competitor, who were under the control of the government - representatives from the Admiralty and Treasury were on the Anglo-Persian board of directors.

In 1919, Anglo-Persian bought up the shares of Broxburn Oil Company and then formed another company to hold these shares. Broxburn Oil Company then operated under its new management company, Scottish Oils. Being its parent company, the Anglo-Persian was basically now the owner of Broxburn Oil Company. They could now run down the company - the profits to Scottish Oils, the losses to Broxburn Oil Company (as the shale and oil workers union had claimed), and then close the company's mines and oil works. Being a government company, the Anglo-Persian was assisted in its objective by the British government, who removed the 6d Excise Duty relief on home-produced oil in 1921, then cancelled the shale-oil companies' Royal Navy contract in 1925.

Despite the fall in oil prices, the British Government's profits from Anglo-Persian for the period 1914-1924 were around $200,000,000 (about £45,000,000) (Winston Churchill: 'The World Crisis'), and during 1924-1925 - $17,500,000 (about £3,800,000), and 1925-1926 - $21,500,000 (about £4,500,000) (Ludwell Denny: 'We Fight For Oil)].

The British government appears to have systematically brought about the closure of the oil works and shale mines at Broxburn. Soon after the official closure was announced in May 1927 the demolition of the works and clearing of the site began. The pulling-down of the biggest oil works in the country appears to have been carried out with indecent haste. There would be no revival for Broxburn Oil Works.

There seems to have been only the desire for immediate profit. No taking into account the possibility of Persia's oil industry being nationalised, or war, both of which would cut off the supply of oil. There was little consideration given to the future, no establishing of strategic oil reserves (stockpiles of oil) to provide economic security during any energy crises which might occur in the future.

Appendices

Norman Macfarlane Henderson

The General Works Manager for the Broxburn Oil Company was Norman Henderson. He had supervised the construction of both Broxburn Refinery and Albyn Oil Works.

He was born in Engine Street, Bathgate, in 1839, eldest child of Alexander Henderson and Elizabeth Liddle, both of whom were cotton hand loom weavers.

His education must have been very basic - at age 11 (now living at 19 Whitburn Road) he was already working as a cotton hand loom weaver. He then became an apprentice joiner and millwright in Bathgate, leaving to take up an engineering apprenticeship with the Shotts Iron Company. In 1861, he became a draughtsman and assistant manager in Young and Meldrum's Bathgate Works.

By 1871 he was already manager of an oilworks (Oakbank Oil Works), and he is listed as living at a place called East Cottage, Midcalder. It was while he was Works Manager at Oakbank that he designed and patented the Henderson retort.

When the Broxburn Oil Company was formed in November 1877, Norman Henderson was appointed Works Manager, and around 1878 he moved into Broxburn Lodge.

The Henderson retort, *opposite*, Patent No 1327, was used in the Broxburn Works. Its chief advantage over previous retorts was that, as well as coal, it burnt what was left of the carbon content (12%) in 'spent' shale as an added source of heat. The retort was large, with a furnace at the bottom into which the 'spent' shale was dropped, then heated. This system resulted in a reduction of 50% in the amount of fuel needed to heat the furnace to the required temperature. Retorts were about 10 feet long or high (horizontal or vertical).

The Henderson retort also removed the need for the awkward water-seal at the bottom which had previously hampered the discharge procedure of the oil, which was now poured out directly into hutches. Shale was tipped from hutches into a hopper (funnel) on top of a retort which was contained within a fire-brick oven. The oven was heated to a high temperature while superheated steam was injected into the top of the retort, extracting the oil from the shale within.

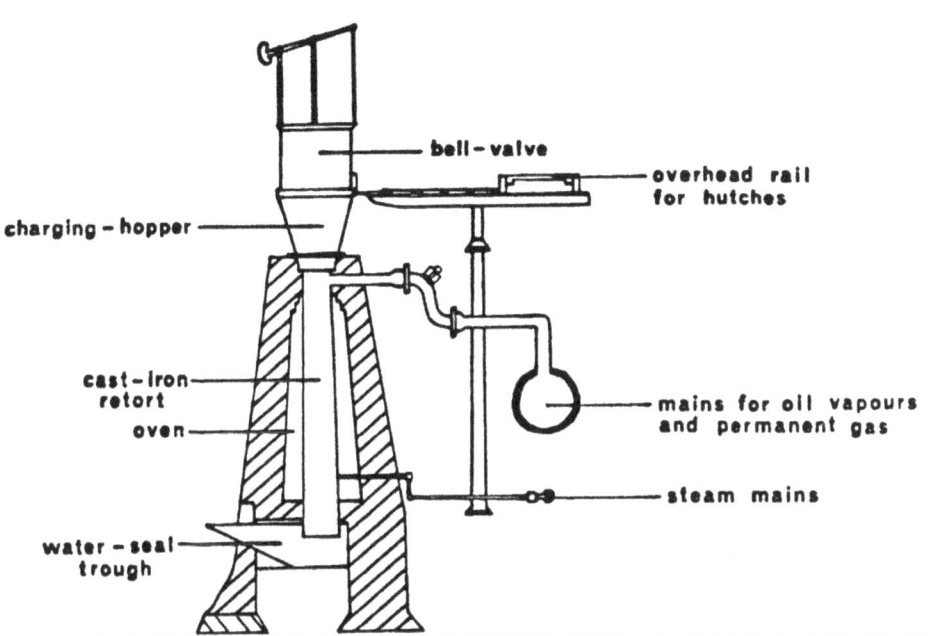

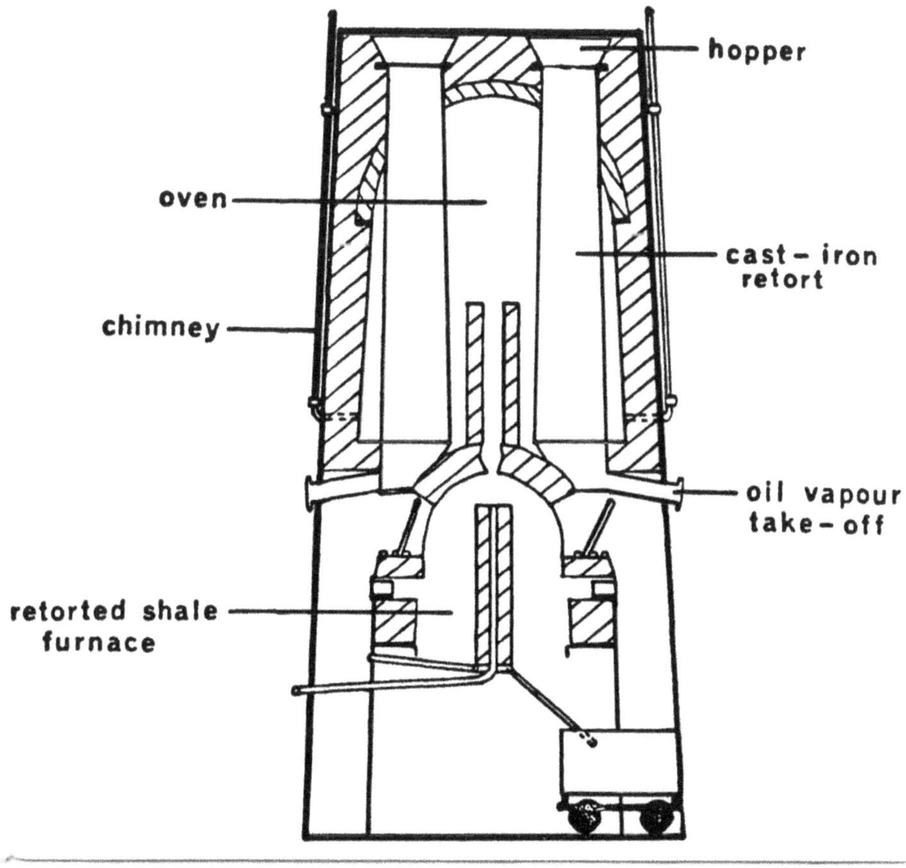

The Henderson Retort

The oil then flowed into a trough below. The Henderson retort also had the oil vapours drawn off at the bottom of the retort, as this was said to 'partially purify them and give a better quality of oil.' The number of workers needed to operate the new retorts was reduced by one third, yet the quantity of oil obtained was greatly increased.

In 1889, Norman Henderson patented another retort (Patent No. 6726), which he named the Broxburn retort. The cast-iron upper section of the new retort was 14 feet high, and the section within the fire-brick oven was 20 feet high. The retorts were constructed in groups of four, each retort having its own flue. Coal and spent shale were heated to produce flammable gas, and this was introduced into the base of the oven while superheated steam was injected. The internal temperature in the retorts was 900° F and in the fire-brick oven, 1300° F.

The capacity was 3.2 tons of crushed shale, and the time taken to process this was 42½ hours. In 24 hours, a bench of 88 retorts could process 160 tons of shale. Per ton of shale, about 23¼ gallons of crude oil, and 11½ lbs of sulphate of ammonia was produced.

Other inventions which Norman Henderson patented were an improved ammonia still, a process for distilling and refining mineral oils, a cooling apparatus for congealing solid paraffin, a method for purifying paraffin wax, and of course his new Broxburn retort.

Norman Henderson became a director of Broxburn Oil Company (mentioned as such in 1902). He also became a J.P. and a County Councillor, and was once a member of Uphall School Board.

In 1911, he was stated to be Consulting Adviser to Broxburn Oil Company.

Norman Macfarlane Henderson died unmarried at Broxburn Lodge, age 78, on 14th December 1917, of locomotor ataxy, nowadays called myasthenia gravis, a chronic disease with progressive muscular paralysis. His occupation is recorded as 'Retired Oil Works Manager'.

The Broxburn Oil Company's Mines to the Dunnet Shale

by
William Clark
Shale Mine Manager at Broxburn
1910

'These mines, situated about a quarter of a mile northeast of Broxburn Village, are only 160 yards distant from the company's shale-breakers and retorts. The motive power chiefly employed is electricity, obtained from a central power-station placed alongside the works boilers and within a few yards of the retorts in which all the exhaust-steam from the engines is utilized: the steam, instead of being supplied directly to the retorts, first passes through the engines, thus saving fuel. The works steam-plant consists of two Stirling boilers and a battery of Lancashire boilers, all equipped with mechanical stokers.

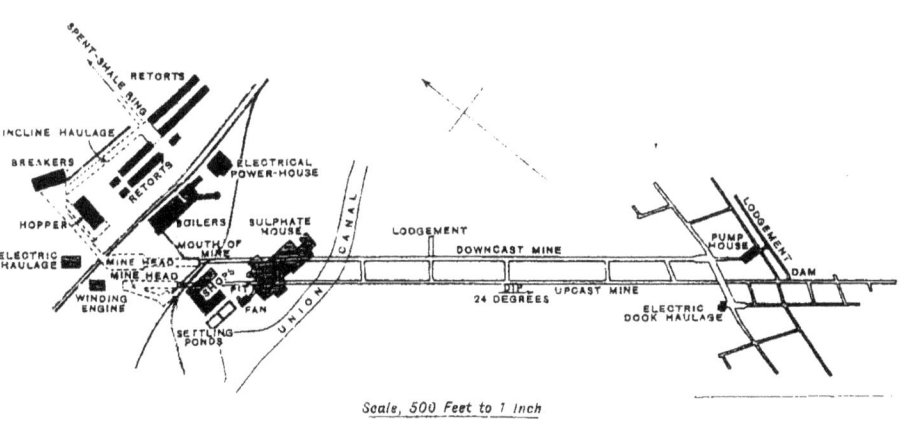

Plan of workings of Dunnet Shale Mine

Sinking - Shale-mines or inclines are usually driven down from the surface of the ground, and follow the seam from the outcrop. The Dunnet Seam, however, does not crop out in the north portion of the company's leasehold, and, in order to reach a large area of it recently disclosed by diamond boring there, these mines have been driven down from the surface at a regular dip. Commencing in the marl immediately below the Broxburn Shale Seam, and passing mostly through fakes and sandstones, they reach the Dunnet Shale Seam at a vertical depth of 720 feet, and at a distance on the slope of over 1,600 feet from the surface.

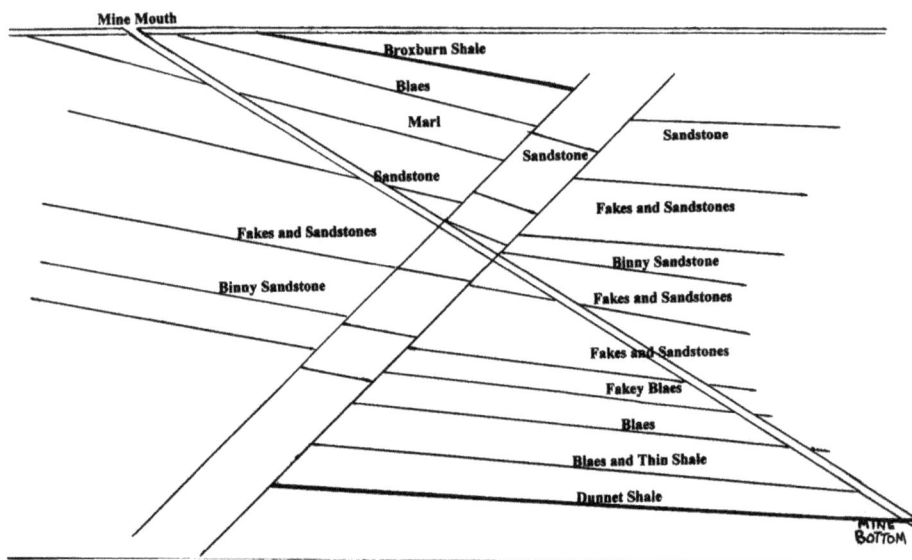

Rock Strata tunnelled through to the Dunnet Shale Mine

Electrical Power House

The Mines, 66 feet apart, are parallel, and connected by "throughers" [passages] at distances varying from 200 to 340 feet. The downcast mine [shaft carrying a current of ventilating air downwards into the mine] is 12 feet wide, and the upcast mine [shaft by which the air current returns from the mine to the surface] 10 feet wide; and both are 7½ feet high. They have brick side-walls throughout, of a minimum thickness of 18 inches, spanned with tram rail girders spaced from 18 in to 4 feet apart.

The girders, which are cleaded [cleated: wood or other material inserted to give additional strength] between with "salted" redwood planks 9 inches broad by 3 inches thick, were tarred before being put into position in order to prevent any possible corrosion due to the "salting" of the wood.

The downcast mine is laid with a double line of rails for endless-rope haulage. A water-track is carried along one side to drain the road and carry away the water given off from several feeders in the sandstones.

This mine is lighted from top to bottom by 32-candle-power incandescent lamps, spaced 120 feet apart.

The upcast mine is used for lowering and raising men and materials. It is laid with a single line of rails of the same section as the downcast mine. The first 300 feet are brick-arched and fitted with two ventilation doors 200 feet apart. In sinking, the boring was done by means of compressed-air drills. The water encountered during sinking amounted to about 130 gallons per minute, and was mostly given off from feeders in the sandstones; it was dealt with by steam-pumps.

Haulage - In the system of haulage adopted, the rope is carried under the hutches, the attachments being made by means of clips which are easily handled by youths.

Endless-Rope Haulage-Gear at the Dunnet Shale Mine

The clips weigh from 18 to 19 pounds. The rope is 4 inches in circumference, and is driven through gearing at a speed of about 1½ miles per hour.

In order to guard against accidents, in the event of a loaded hutch breaking loose and running back down the up-track, the following invention has been provided for automatically transmitting a signal or effecting a stoppage of the haulage, or both – the invention consists of a contact device comprising of a pivoted arm arranged to project into the pathway of the axle of the hutch. This arm projects upwards from between the rails, and while it is adapted to swing wildly when passed by a hutch going up the incline, if acted upon by a hutch coming down the track, the arm is forced back and completes an electrical circuit for releasing gear so as to stop the engine. The apparatus has been very thoroughly tested both on the surface and in the mine, and has yielded most satisfactory results. In the upcast mine, the raising and lowering of men and material is done by a pair of coupled winding steam-engines.

Mine-head - The loaded hutches are delivered from the main haulage on the mine-head, and after passing over the weighing-machines, are carried to the breaking machines; here they are emptied by an automatic revolving tippler; they are then raised and gravitated back to the head of the haulage which delivers them at the mine-head, ready for returning down the mine. A bunker for storing broken shale, with a capacity of 2,000 tons, has been erected alongside the breakers.

Ventilation - For ventilation purposes an 18-foot fan and engine is provided. Steam for this engine is got from the sulphate-house boilers, the exhaust-steam being passed back to the sulphate-house and utilized there.

System of Working - The seam is 6 feet thick, overlain by 3 feet of shaly Blaes [clay which contains little or no shale] The system of working is stoop and room. A dook [an inclined passage] is also being driven down. The main bottom is 16 feet wide, with brick side-walls, and roofed with 10 by 6-inch girders of H section, cleaded with "salted" redwood planks. It is arranged so that the loaded hutches will gravitate from the top of the dook to the foot of the downcast mine. A motor-generator is being installed underground capable of supplying power to ten electric drills, to be utilised for boring shot-holes in the shale.'

Dunnet Shale Mine beside Albyn Crude Oil Works, 1910.

Description of the Broxburn Works

by
William Love
Managing Director of Broxburn Oil Company
1910

The works and mines now cover upwards of 250 acres, and give employment to about 1,800 workpeople, of whom one-half are employed in the mines. About 1,600 tons of oil-shale daily, or 500,000 tons per year, are distilled, and some 18,000,000 gallons of crude oil refined per annum. The consumption of coal averages 400 tons a day.

The Broxburn Works comprise crude-oil works, sulphate house, refinery, acid works, candle manufactory, and cooperage. The Roman Camp Works, which manufactures crude oil only, are about 1½ miles distant, and are connected with the refinery by a line of railway.

The mineral field has an area of about 3,000 acres, and the oil-shale is mined in the same manner as coal. The field is partly in the form of a dome, the crude-oil works being placed in the centre, from which the mines radiate, and pits are sunk to the deeper parts of the shale-seams for the drainage of the workings. Mines are also put down on other anticlines.

At the mines farthest away from the crude-oil works, the miners' hutches, on reaching the surface are emptied into railway-wagons which are conveyed to the hydraulic tippers, and the shale shot into hoppers leading to the breaker. At the mines nearer the works, the hutches are taken direct to the breaker. The broken shale from the breakers falls into hutches carrying about 17 hundredweights each, and is conveyed by an endless-rope haulage to the platform at the top of the various benches of retorts. The hutches are discharged into iron hoppers holding about 30 hours' supply of shale which is mechanically fed into the retorts beneath, and 4½ tons are put through each retort per day.

There are 240 Henderson vertical retorts in use at the Broxburn Works, and 160 at the Roman Camp Works, or 400 retorts in all. The retorts are about 34 feet long, the upper portion being of cast-iron (some 14 feet long) and the lower portion of brick. The oil is distilled from the shale in the cast-iron part at a temperature of about 1,100° Fahr. in the oven, and the spent shale is passed downwards into the brickwork portion of the retort which is kept at a temperature of from 1,600° to 1,800° Fahr. in the flue. Steam is passed in at the bottom of the retort, and, becoming superheated, reacts on the carbon to produce water-gas, and on the nitrogen to form ammonia. The high temperature increases the yield of ammonia and permanent gases from the carbon of the shale. The products of distillation are withdrawn at the top of the retorts by a branch pipe. In passing downwards through the retorts, the shale is kept in constant motion by means of toothed rollers at the bottom, which deliver the spent shale into a hopper. From this hopper the spent shale is discharged into steel hutches holding about 25 hundredweights each. These hutches are conveyed to the spent-shale bing by an endless-rope haulage. The retorts are heated by incondensable gas generated by the retorts themselves, introduced by ports at the sides.

The Broxburn shales yield 22 gallons of crude oil and 40 pounds of sulphate of ammonia per ton; while the Roman Camp shales yield 20 gallons of crude oil and about 60 pounds of sulphate of ammonia per ton.

The products of distillation in the form of oil and water-vapour, together with the incondensable gases, are drawn by gas-exhausters from the top of the retorts into a large main, and through a series of vertical cooling-pipes and scrubbing towers, where the crude oil and ammonia water are condensed. The liquids flow into a separator-tank, and a pipe fitted close to the bottom draws off the ammonia-water, while another pipe fitted near the top of the tank draws off the crude oil, each into its own receptacle. The incondensable or permanent gases are returned to the retorts as fuel.

In the manufacture of sulphate of ammonia, the ammonia-liquor is passed into a columnar still provided with a series of horizontal trays. The liquor is admitted at the top, and flows from tray to tray until it is discharged at the bottom of the still, where steam is introduced at a pressure of about 50 pounds per square inch. The steam, ammonia, and other gases, escape at the top of the still, and are conveyed through lead pipes to the bottom of a saturator-box. The saturator-box is fed with sulphuric acid, through which the gases bubble from perforations in the lead pipe. The ammonia forms sulphate of ammonia, whilst steam, carbon dioxide, and sulphuretted hydrogen are returned from the saturator-box to the retorts. The sulphate of ammonia is lifted out of the saturator-box, as it is formed, by means of steam-injectors. The salt is drained for about 12 hours; it is then conveyed into the store-house, and after drying for a few days is passed through a crushing-mill and placed in bags. The recovered acid is saturated with ammonia in the same way, but produces a solution of sulphate of ammonia. The water is evaporated by exhaust-steam in a vacuum apparatus. The salt is dried by a centrifugal machine, and conveyed to the store by mechanical conveyors.

The crude oil is pumped into a large stock-tank in the refinery. It is afterwards pumped into charging-tanks, from which it flows, by gravity, to two sets of two-crude-oil stills, coupled together by pipes, and worked on the Henderson continuous system. In the subsequent distillations this continuous method comes even more into play.

The first distillation of the crude oil yields green naphtha, green oil, and still-coke (a valuable smokeless fuel). The green naphtha is treated with sulphuric acid and caustic soda, and, after distillation by steam, yields shale-spirit or naphtha of a specific gravity from 0.710 to 0.750. The green oil is also treated with sulphuric acid and caustic soda, and is then ready for the second distillation. In these treatments the oil is stirred with sulphuric acid and allowed to settle, when a black tar falls to the bottom of the tank and is run off, while the oil is run by gravitation from the upper to the lower washer, where it is mixed with a solution of caustic soda, and on settling, another black tar separates. From these tars the sulphuric acid is removed by washing, and is used in the manufacture of sulphate of ammonia. The washed tars are injected by steam-jets into the still-furnaces in the form of fine spray, and burned as fuel.

By a second distillation, the green oil is fractionated into light and heavy oils, the latter containing solid paraffin. The light oils are treated and distilled once, and a small portion of it twice, for the following brands of burning oils:-

Petroline
No. 1 paraffin oil
Lighthouse oil
Marine sperm

In coking in the residue stills, much gas is evolved, which used to escape into the atmosphere. Recently, Mr. Henderson patented a process for the recovery of this gas, which is very pure and rich, and used for lighting and heating, and is thus a valuable bye-product. Light naphtha is got from it also, and used for motor spirit.

To obtain the solid paraffin from the heavy oil, it is cooled down below 32° Fahr., and the paraffin strained out by means of filter-presses, the products being blue oil and crude solid paraffin. The blue oil is treated, distilled, and fractionated into intermediate oils used for gas-making, fuel, and other purposes, and heavy oils used for lubricating machinery. The crude solid paraffin is passed to the sweating-house, where it is fractionated and refined by the Henderson process to a melting point of 120° Fahr., and is known as semi-refined paraffin wax. The sweating and fractionating process is repeated to produce refined paraffin-wax.

After the sweating process, the hot liquid paraffin is mixed with char, settled, and filtered through paper. It is then cooled into blocks of paraffin-wax. About 6,000 tons of refined paraffin-wax is produced per annum, the greater part of which is made into candles before leaving the works.

The candle factory is capable of producing about 20 tons of candles per day, and is fully equipped with the most modern forms of packing-case nailing machinery.

Paraffin-wax is without taste, smell, or colour, and resists the action of the strongest acids. The harder waxes are used as illuminants in the form of candles, night-lights, tapers, and wax matches; and the softer waxes are burnt in cera-lamps, miners' safety-lamps, and in hand-lamps.

The soft 100° Fahr. wax, or match-paraffin, is used instead of sulphur to saturate the sticks of wooden matches, so as to enable them to ignite readily.

The sulphuric-acid manufacturing plant is divided into two works, one consisting of five chambers, two Glover towers, and a Gay-Lussac tower, while the other consists of two sets of three chambers, each set having a Glover tower and a Gay-Lussac tower. One plant is provided with a fan to assist the draught; the other has chimney draught only.

There is a Kessler concentrating plant, capable of making 14 to 16 tons of strong acid per day.

The cooperage comprises a drying-shed for barrels; coopers' shop for repairing barrels; a glue-house, where the insides of the barrels are covered with a thin coating of glue, so as to close the seams and prevent the oil from escaping through the pores of the wood; a shop for painting the barrels; and the filling-house, fitted with the Crawford filling apparatus, which automatically shuts off the oil when the barrel is full.'

Shale Oil Works Chronology

Greendykes | Stewartfield

Greendykes	Stewartfield
1862 Oil works of Broxburn Shale Oil Co Ltd (Robert Bell) Oil works of James Steele	
1863 Oil works of Broxburn Shale Oil Co Ltd (Robert Bell) Oil works of James Steele	Oil works of Robert Bell
1864 Oil works of Edward Fernie Oil works of James Steele Oil refinery of Thomas Hutchison	Oil works of Robert Bell
1865 Oil works of Ebenezer Waugh Fernie Oil works of James Steele Oil refinery of Thomas Hutchison Oil works of Robert Bell	Oil works of Robert Bell Oil works of James Steele
Oil refinery of James Vallance at the Arches	

1866	
Albyn Oil Works of Glasgow Oil Company Broxburn Ltd Buchan Oil Works of John Poynter Oil refinery of Thomas Hutchison Oil works of Robert Bell Oil works of James Miller	Oil works of Robert Bell Oil works of James Steele

Oil refinery of James Vallance at the Arches

1867	
Albyn Oil Works of Glasgow Oil Company Broxburn Ltd Buchan Oil Works of John Poynter Oil refinery of Thomas Hutchison Oil works of Robert Bell Oil works of James Miller	Oil works of Robert Bell Oil works of James Steele

Oil refinery of James Vallance at the Arches

1868	
Albyn Oil Works of Glasgow Oil Company Broxburn Ltd Buchan Oil Works of John Poynter Oil works of Robert Bell Oil works of James Miller	Oil works of Robert Bell

Oil refinery of James Vallance at the Arches

1869 Albyn Oil Works of Glasgow Oil Company Broxburn Ltd Buchan Oil Works of John Poynter Oil works of Robert Bell Oil works of James Miller Oil works of James Liddell	Oil works of Robert Bell

Oil refinery of James Vallance at the Arches

1870 Albyn Oil Works of Glasgow Oil Company Broxburn Ltd Buchan Oil Works of John Poynter Oil works of Robert Bell Oil works of James Miller Oil works of James Liddell	Oil works of Robert Bell

Oil refinery of James Vallance at the Arches

1871 Albyn Oil Works of Glasgow Oil Company Broxburn Ltd Buchan Oil Works of John Poynter Oil works of Robert Bell Oil works of James Miller Oil works of James Liddell	Oil works of Robert Bell

1872 Albyn Oil Works of Glasgow Oil Company Broxburn Ltd Oil works of Robert Bell Oil works of James Liddell Oil refinery of Robert Bell	Oil works of Robert Bell
1873 Albyn Oil Works of Glasgow Oil Company Broxburn Ltd Oil works of Robert Bell Oil works of James Liddell Oil refinery of Robert Bell	Oil works of Robert Bell
1874 Albyn Oil Works of Glasgow Oil Company Broxburn Ltd Oil works of Robert Bell Oil works of James Liddell Oil refinery of Robert Bell	Oil works of Robert Bell

1875 Albyn Oil Works of Glasgow Oil Company Broxburn Ltd Oil works of Robert Bell Oil works of James Liddell & Company Oil refinery of Robert Bell	Oil works of Robert Bell

1876 Albyn Oil Works of Glasgow Oil Company Broxburn Ltd Oil works of Robert Bell Oil works of James Liddell & Company Oil refinery of Robert Bell Oil works of James Miller	

1877 Oil works of Broxburn Oil Company Ltd Oil refinery of Broxburn Oil Company Ltd	

Oil works of George Simpson at East Mains

1878 Albyn Oil Works of Broxburn Oil Company Ltd Oil refinery of Broxburn Oil Company Ltd [Broxburn Oil Works]	

Oil works of George Simpson at East Mains	
1879-1919 Albyn Oil Works of Broxburn Oil Company Ltd Oil refinery of Broxburn Oil Company Ltd [Broxburn Oil Works]	
1919-1927 Albyn Oil Works of Scottish Oils Limited Broxburn Oil Works of Scottish Oils Limited	

Early Management and Workforce of Broxburn Oil Company

Chairmen:
1877 - 1887 - William Kennedy, Glasgow.
1887 - 1894 - Robert Bell.
1894 - c1904 - Sir James Steel, Lord Provost of Edinburgh.
c1904 - c1907 - Sir David Richmond, Glasgow.
c1907 - c1909 - James Stedman Dixon, Glasgow.
c1909 - Sir Archibald McInnes Shaw, Lord Provost of Glasgow.
[Still Chairman in 1911]

Managing Directors
1877 - 1900 - William Kennedy, Glasgow.
1900 - William Love, Saltcoats.
[Still Managing Director in 1911]

1879 - The number of workers employed by Broxburn Oil Company was 850.
1881 - Broxburn Oil Company employed 806 men and 63 boys.
1882 - The workforce was 1200. (800 miners, and 400 oil workers.)
1896 - In their mines at Broxburn, the company employed 463 underground workers and 49 surface workers. The manager was William Clark. (Presumably he was Mine Manager, since Norman Henderson was Works Manager of Broxburn Oil Company.)
1902 - The number of Broxburn Oil Company employees was 1850.
1908 - The number of employees in the shale mines at Broxburn was 521 underground workers and 100 surface workers. The manager was William Clark. (Mine Manager). His under manager was W. Murray.

Broxburn Oil Company Shale Output, 1877 - 1925

Payments to Lord Cardross, later Earl of Buchan from 1898

	Shale Tons - Cwts	per Ton	£	s	d
12 Nov 1877-15 May 1878 -	46,268 18	9d	1,735	1	5
16 May 1878-11 Nov 1878 -	44,197 4		1,657	7	7
12 Nov 1878-15 May 1879 -	77,051 11		2,889	8	6
16 May 1879-11 Nov 1879 -	87,857 13		3,294	13	1
12 Nov 1879-15 May 1880 -	96,569 8		3,621	7	1
16 May 1880-11 Nov 1880 -	99,873 0		3,745	4	9
12 Nov 1880-14 May 1881 -	99,383 11		3,726	17	7
16 May 1881-11 Nov 1881 -	124,876 3		4,682	17	0
12 Nov 1881-15 May 1882 -	122,496 10		4,593	12	4
16 May 1882-11 Nov 1882 -	111,732 6		4,190	19	0
12 Nov 1882-15 May 1883 -	128,190 17		4,807	3	1
6 May 1883-10 Nov 1883 -	107,601 11		4,035	1	2
12 Nov 1883-15 May 1884 -	130,004 19		4,875	3	9
16 May 1884-11 Nov 1884 -	147,848 2		5,544	6	0
11 Nov 1884-15 May 1885 -	163,153 12		6,118	4	2
16 May 1885-11 Nov 1885 -	159,002 13		5,962	12	0
12 Nov 1885-15 May 1886 -	155,557 9		5,833	8	1
16 May 1886-11 Nov 1886 -	138,780 14		5,204	5	6
12 Nov 1886-14 May 1887 -	42,784 19		5,354	8	8
16 May 1887-11 Nov 1887					
Ordinary Shale	45,340 5	9d	1,700	5	2
Opencast Shale	6,410 9	6d	160	5	2
Total:	51,750 14		1,860	10	4

12 Nov 1887-15 May 1888				
Ordinary shale	156,433 16		5,866 5 4	
Opencast shale	10,586 9		264 13 2	
Total:	167,020 5		6,130 18 6	
16 May 1888-10 Nov 1888				
Ordinary shale	181,284 11		6,798 3 5	
Opencast shale	188 17		4 14 5	
Total:	181,473 8		6,802 17 10	
12 Nov 1888-15 May 1889				
Ordinary shale	178,876 7		6,707 17 3	
16 May 1889-11 Nov 1889				
Ordinary shale	170,417 0		6,390 12 0	
Total:	349,293 7		13,098 10 0	
11 Nov 1889-15 May 1890				
Ordinary shale	169,199 18		6,344 19 11	
Opencast shale	1,161 16	3d	14 10 5	
Total:	170,361 14		6,359 10 4	
16 May 1890-11 Nov 1890				
Ordinary shale	160,478 16		6,017 19 1	
Opencast shale	478 7	6d	11 19 2	
Total:	160,957 3		6,029 18 3	
11 Nov 1890-15 May 1891				
Ordinary shale	164,859 13		6,182 4 9	
16 May 1891-11 Nov 1891				
Ordinary shale	159,125 14		5,967 4 3	
Roman Camp shale	4,841 0	6d	121 0 6	
Total:	163,966 14		6,088 4 9	
12 Nov 1891-15 May 1892				
Ordinary shale	154,536 16		5,795 2 7	
Roman Camp shale	13,961 12		349 0 10	
Total:	168,498 8		6,144 3 5	
16 May 1892-11 Nov 1892				
Ordinary shale	119,948 15		4,498 1 6	
Roman Camp shale	2,568 12		64 4 3	
Total:	122,517 7		4,562 5 9	
12 Nov 1892-15 May 1893				
Ordinary shale	123,881 3		4,645 10 10	
Roman Camp shale	10,738 19		268 9 5	
Total:	134,620 2		4,914 0 3	

16 May 1893-11 Nov 1893
- Ordinary shale   76,522 3   2,869 11 7
- Roman Camp shale   48,877 12   1,221 18 10
- Total: 125,399 15   4,091 10 5

13 Nov 1893-15 May 1894
- Ordinary shale   101,204 2   6d   2,530 2 0
- Roman Camp shale   66,720 15   6d   1,668 0 4
- Total: 167,924 17   4,198 2 4

16 May 1894-11 Nov 1894
- Ordinary shale   90,568 7   2,264 4 2
- Roman Camp shale   63,748 16   1,593 14 5
- Total: 154,317 3   3,857 18 7

11 Nov 1894-15 May 1895
- Ordinary shale   119,412 17   2,985 6 5
- Roman Camp shale   67,024 11   1,675 12 3
- Total: 186,437 8   4,660 18 8

16 May 1895-11 Nov 1895
- Ordinary shale   125,316 4   3,132 18 2
- Roman Camp shale   64,550 11   1,613 15 2
- Total: 189,866 15   4,746 13 4

12 Nov 1895-15 May 1896
- Ordinary shale   130,361 15   3,259 0 10
- Roman Camp shale   67,299 15   1,682 9 10
- Total: 197,661 10   4,941 10 8

16 May 1896-11 Nov 1896
- Ordinary shale   128,061 16   3,201 10 10
- Roman Camp shale   67,557 16   1,688 18 10
- Total: 195,619 12   4,890 9 8

12 Nov 1896-15 May 1897
- Ordinary shale   147,794 6   3,694 17 1
- Roman Camp shale   71,067 14   1,776 13 10
- Total: 218,862 0   5,471 10 11

15 May 1897-11 Nov 1897
- Ordinary shale   145,916 6   6d   3,647 18 1
- Roman Camp shale   70,305 2   6⅔d   1,952 18 5
- Total: 216,221 8   5,600 16 6

12 Nov 1897-15 May 1898
    Ordinary shale   167,997 15   4,199 18 8
    Roman Camp shale   72,313 7½   2,008 13 8
    Total: 240,311 2½   6,208 12 4

16 May 1898-11 Nov 1898
    Ordinary shale   148,082 4   3,702 1 0
    Roman Camp shale   60,470 17   1,679 14 6
    Total: 208,553 1   5,381 15 6

12 Nov 1898-15 May 1899
    Ordinary shale   189,950 0   4,748 15 0
    Roman Camp shale   74,795 4½   2,077 12 10
    Total: 264,745 4½   6,826 7 10

15 May 1899-11 Nov 1899
    Ordinary shale   163,558 14   4,088 19 5
    Roman Camp shale   72,251 7   2,006 19 11
    Total: 235,810 1   6,095 19 4

13 Nov 1899-15 May 1900
    Ordinary shale   174,714 6   4,367 17 2
    Roman Camp shale   74,306 6   2,064 1 3
    Total: 249,020 12   6,431 18 5

15 May 1900-11 Nov 1900
    Ordinary shale   39,617 2   10d   1,650 14 1
    Roman Camp shale   74,419 19   6⅔d   2,067 4 4
    Curly shale   127,123 7   6d   3,178 7 8
    Total: 241,160 8   6,896 6 1

12 Nov 1900-15 May 1901
    Ordinary shale   36,628 19   1,526 4 1
    Roman Camp shale   78.625 12   2,184 0 10
    Curly shale   142,504 3   3,562 12 1
    Total: 257,758 14   7,272 17 0

16 May 1901-11 Nov 1901
    Ordinary shale   26,150 0   1,089 11 8
    Roman Camp shale   70,624 16   1,961 15 10
    Curly shale   134,184 17   3,354 12 4
    Total: 230,959 13   6,405 19 10

12 Nov 1901-15 May 1902
    Ordinary shale   20,318 6   846 11 11
    Roman Camp shale   72,205 0   2,005 13 10

Curly shale	143,748 9		3,593 14 3
Grey shale	6,527 5	3¼d	88 7 8
Total:	242,799 0		6,534 7 8

16 May 1902-11 Nov 1902
Ordinary shale	15,559 17	648 6 6
Roman Camp shale	67,642 9	1,878 19 1
Curly shale	140,041 15	3,501 0 10
Grey shale	16,277 8	220 8 4
Total:	239,521 9	6,248 14 9

12 Nov 1902-15 May 1903
Ordinary shale	9,920 5	413 6 10
Roman Camp shale	69,914 5	1,942 1 0
Curly shale	147,000 1	3,675 0 0
Grey shale	16,406 15	222 3 5
Total:	243,241 6	6,252 11 3

16 May 1903-11 Nov 1903
Ordinary shale	5,887 4	245 6 0
Roman Camp shale	68,155 16	1,893 4 4
Curly shale	131,815 13	3.295 7 9
Grey shale	31,376 13	424 17 9
Total:	237,235 6	5,858 15 10

12 Nov 1903-15 May 1904
Ordinary shale	3,548 6	147 16 11
Roman Camp shale	72,041 11	2,001 3 0
Curly shale	136,546 16	3,413 13 4
Grey shale	39,311 5	532 6 6
Total:	251,447 18	6,094 19 9

16 May 1904-11 Nov 1904
Ordinary shale	3,396 14	141 10 7
Roman Camp shale	76,742 1	2,131 14 2
Curly shale	125,946 17	3,148 13 4
Grey shale	63,217 19	856 1 4
Total:	269,303 11	6,277 19 5

12 Nov 1904-15 May 1905
Ordinary shale	5,257 17	219 1 6
Roman Camp shale	74,303 13	2.063 19 5
Curly shale	115,294 7	2,882 7 2
Grey shale	74,415 11	1,007 14 1
Total:	269,271 8	6,173 2 2

16 May 1905-11 Nov 1905
    Ordinary shale   9,634 13   401 8 10
    Roman Camp shale   92,772 11   2,577 0 3
    Curly shale 97,958 6   2,448 19 1
    Grey shale 69,962 18   947 8 2
Total: 270,328 8   6,374 16 4

13 Nov 1905-15 May 1906
    Ordinary shale   11,254 8   468 18 8
    Roman Camp shale   88,535 13   2,459 6 5
    Curly shale   99,897 13   2,497 8 10
    Grey shale   74,600 17   1,010 4 4
    Blaes   6,743 19   1d   28 1 11
Total: 281,032 10   6,464 0 2

16 May 1906-10 Nov 1906
    Ordinary shale   8,123 13   338 9 8
    Roman Camp shale   83,266 8   2,312 19 1
    Curly shale   94,950 18   2,373 15 5
    Grey shale   64,361 7   871 11 2
    Blaes   9,432 2   39 6 0
Total: 260,134 8   5,936 1 4

12 Nov 1906-15 May 1907
    Ordinary shale   12,216 5   509 0 2
    Roman Camp shale   89,265 0   2,479 11 8
    Curly shale   99,754 8   2,493 17 2
    Grey shale   63,367 19   858 2 1
    Blaes   10,053 0   41 17 9
Total: 274,656 12   6,382 8 10

16 May 1907-11 Nov 1907
    Ordinary shale   17,824 10   742 13 9
    Roman Camp shale   85,669 15   2,379 14 3
    Curly shale   91,168 7   2,279 4 2
    Grey shale   52,492 6   710 16 8
    Blaes   11,293 4   47 1 1
Total: 258,448 2   6,159 9 11

12 Nov 1907-15 May 1908
    Ordinary shale   17,108 11   712 17 1
    Roman Camp shale   90,219 4   2,506 1 9
    Curly shale   105,046 14   2,626 3 4
    Grey shale   54,129 8   733 0 0
    Blaes   10,963 16   45 13 7
Total: 277,467 13   6,623 15 9

16 May 1908-11 Nov 1908			
Ordinary shale	17,199 14		716 13 1
Roman Camp shale	83,443 3		2,317 17 3
Curly shale	89,509 8		2,237 14 8
Grey shale	67,959 6		920 5 7
Blaes	10,852 4		45 4 4
Total:	268,963 15		6,237 14 11
12 Nov 1908-15 May 1909			
Ordinary shale	17,995 13		749 16 4
Roman Camp shale	87,122 5		2,420 1 3
Curly shale	78,323 4		1,958 1 7
Grey shale	94,030 2		1,273 6 6
Blaes	12,781 19		53 5 2
Total:	290,253 3		6,454 10 10
17 May 1909- 11 Nov 1909			
Ordinary shale	15,912 3		663 0 1
Roman Camp shale	83,811 1		2,328 1 8
Curly shale	65,650 13		1,641 5 3
Grey shale	95,760 13		1,296 15 2
Blaes	12,347 1		51 8 11
Total:	273,481 11		5,980 11 1
12 Nov 1909-15 May 1910			
Ordinary shale	24,766 15		1,031 18 11
Roman Camp shale	82,985 10		2,305 3 0
Curly shale	56,296 18		1,407 8 5
Grey shale	91,896 17		1,244 8 8
Upper Grey shale	7,188 13	3¼d	97 6 11
Dunnet shale	6,389 5	4½d	119 15 10
Blaes	13,420 19		55 18 4
Total:	282,944 17		6,262 0 1
16 May 1910-11 Nov 1910			
Ordinary shale	30,611 17		1,275 9 10
Roman Camp shale	84,114 14		2,336 10 4
Curly shale	52,732 15		1,318 6 4
Grey shale	73,380 16		993 13 11
Dunnet shale	17,184 6		322 4 1
Blaes	20,365 17		84 17 1
Total:	278,390 5		6,331 1 7

12 Nov 1910-15 May 1911
Ordinary shale	20,880 3		870 0 1
Roman Camp shale	96,756 0		2,687 13 4
Curly shale	46,130 19		1,153 5 6
Grey shale	91,455 11		1,238 9 2
Dunnet shale	31,218 2		585 6 9
Blaes	14,800 19		61 13 5
Total:	301,241 14		6,596 8 3

16 May 1911-11 Nov 1911
Ordinary shale	16,961 14		706 14 9
Roman Camp shale	89,664 0		2,490 13 4
Curly shale	38,469 13		961 14 10
Grey shale	95,568 19		1,294 3 3
Dunnet shale	47,167 11		884 7 10
Blaes	13,498 4		56 4 10
Total;	301,330 1		6,393 18 10

12 Nov 1911-15 May 1912
Ordinary shale	17,296 7		720 13 7
Roman Camp shale	85,761 12		2,382 5 4
Curly shale	48,478 4		1,211 19 1
Grey shale	87,407 14		1,183 12 11
Upper Grey shale	265 14	3¼d	3 11 11
Dunnet shale	63,438 15		1,189 9 6
Blaes	17,348 13		72 5 8
Total:	319,996 19		6,763 18 0

16 May 1912-11 Nov 1912
Ordinary shale	20,824 11		867 13 9
Roman Camp shale	82,661 18		2,296 3 3
Curly shale	38,182 11		954 11 3
Grey shale	79,414 10		1,075 8 1
Dunnet shale	63,704 1		1,194 9 1
Blaes	22,678 14		94 9 10
Total:	307,466 5		6,482 15 3

12 Nov 1912-15 May 1913
Ordinary shale	18,843 4		785 2 8
Roman Camp shale	88,633 19		2,462 1 1
Curly shale	41,519 9		1,037 19 9
Grey shale	83,426 19		1,129 14 10
Upper Grey shale	31 6		0 8 6
Dunnet shale	66,610 5		1,248 18 10
Blaes	24,232 6		100 19 4
Total:	323,297 8		6,765 5 0

16 May 1913-11 Nov 1913		
Ordinary shale	19,122 6	796 15 3
Roman Camp shale	83,968 7	2,332 9 1
Curly shale	29,437 17	735 18 9
Grey shale	83,226 15	1,127 0 6
Upper Grey shale	1,899 5	25 14 4
Dunnet shale	65,300 19	1,224 7 9
Blaes	25,327 18	105 10 7
Total:	308,283 7	6,347 16 3
12 Nov 1913-15 May 1914		
Ordinary shale	24,485 16	1,020 4 10
Roman Camp shale	92,667 3	2,574 1 8
Curly shale	16,689 8	417 4 9
Grey shale	95,706 17	1,296 0 7
Upper Grey shale	2,317 11	31 7 8
Dunnet shale	66,005 4	1,237 12 0
Blaes	27,830 2	115 19 2
Total:	325,702 1	6,692 10 8
16 May 1914-11 Nov 1914		
Ordinary shale	7,091 3	295 9 3
Roman Camp shale	86,313 19	2,397 12 2
Curly shale	14,715 5	367 17 8
Grey shale	104,370 10	1,413 7 0
Upper Grey shale	1,236 14	16 14 11
Dunnet shale	69,206 2	1,297 12 3
Blaes	18,329 16	76 7 6
Total:	301,263 9	5,865 0 9
12 Nov 1914-15 May 1915		
Ordinary shale	7,709 15	321 4 9½
Roman Camp shale	89,960 18	2,498 18 2
Curly shale	33,889 4	847 4 7
Grey shale	83,712 19	1,133 12 3
Dunnet shale	69,753 18	1,307 17 8
Blaes	20,652 0	86 1 0
Total:	305,678 14	6,194 18 5½
16 May 1915-11 Nov 1915		
Ordinary shale	8,955 17	373 3 2
Roman Camp shale	94,704 10	2,630 13 7
Curly shale	55,217 10	1,380 8 9
Grey shale	39,961 6	541 2 9
Dunnet shale	70,954 3	1,330 7 9
Blaes	20,346 17	84 15 6
Total:	290,140 3	6,340 11 6

12 Nov 1915-15 May 1916
Ordinary shale	16,016 5		667 6	10
Roman Camp shale	97,870 13		2,718 12	7
Curly shale	54,004 4		1,350 2	1
Grey shale	12,339 4		167 1	10½
Dunnet shale	80,432 19		1,508 2	4½
Blaes	26,090 10		108 14	2
Total:	286,753 15		6,519 19	11

16 May 1916-11 Nov 1916
Ordinary shale	16,006 16		666 19	0
Roman Camp shale	94,092 12		2,613 13	8
Curly shale	57,266 9		1,431 13	2
Grey shale	16,192 8		219 5	5
Dunnet shale	77,850 1		1,459 13	9
Blaes	26,452 7		110 4	4
Total:	287,860 13		6,501 9	4

12 Nov 1916-15 May 1917
Roman Camp shale	93,753 2		2,604 3	0½
Curly shale	58,376 2		1,459 8	0
Grey shale	26,896 9		364 4	5
Dunnet shale	83,838 8		1,571 19	5
Blaes	28,510 1		118 15	10
Total:	291,374 2		6,118 10	8½

16 May 1917-10 Nov 1917
Ordinary shale	18,048 8		752 0	4
Roman Camp shale	86,765 8		2,410 3	0
Curly shale	52,380 10		1,309 10	3
Grey shale	30,458 8		412 9	2
Dunnet shale	83,406 0		1,563 17	3
Blaes	28,832 18		120 2	9
Camps shale	296 11	4¼d	5 4	10
Raeburn shale	160 0	4d	2 13	4
Total:	300,348 3		6,576 0	11

12 Nov 1917-15 May 1918
Ordinary shale	15,503 0	10d	645 19	2
Roman Camp shale	86,690 13	6⅔d	2,408 1	5
Curly shale	54,288 9	6d	1,357 4	3
Grey shale	26,059 5	3¼d	352 17	8
Upper Grey shale	2,132 3	3¼d	28 17	5
Dunnet shale	87,299 3	4½d	1,636 17	1½
Blaes	28,011 17	1d	116 14	4
Camps shale	637 12	4¼d	11 5	9
Raeburn shale	901 17	4d	15 0	7
Total:	301,523 19		6,572 17	8½

16 May 1918-11 Nov 1918

Ordinary shale	14,086 15		586 18 11½
Roman Camp shale	81,905 1		2,275 2 9
Curly shale	51,319 12		1,282 19 9½
Grey shale	24,635 10		333 12 1¼
Upper Grey shale	2,355 6		31 17 11
Dunnet shale	86,137 6		1,615 1 5½
Blaes	22,832 6		95 2 8
Camps shale	1,338 18		23 14 2
Raeburn shale	321 19		5 7 4
Total:	284,932 13		6,249 17 1¾

12 Nov 1918-15 May 1919

Ordinary shale	20,000 9		833 11 2½
Roman Camp shale	82,891 13		2,302 10 11
Curly shale	41,505 10		1,037 12 9
Grey shale	28,036 17		379 13 3¾
Upper Grey shale	3,706 3		50 3 9
Dunnet shale	87,266 11		1,636 4 11½
Blaes	22,656 14		94 8 0½
Camps shale	1,621 11		28 14 3½
Raeburn shale	495 17		8 5 3½
Total:	288,181 5		6,371 4 6¼

16 May 1919-11 Nov 1919

Ordinary shale	15,191 6		632 19 5
Roman Camp shale	68,069 14		1,890 16 1
Curly shale	32,526 15		813 3 4½
Grey shale	14,496 4		196 6 0½
Upper Grey shale	2,925 0		39 12 2¼
Dunnet shale	67,806 7		1,271 7 4½
Blaes	16,077 6		66 19 9¼
Raeburn shale	1,003 5		16 14 5
Barracks shale	1,298 15	4¼d	22 19 11½
Total:	219,394 12		4,950 18 7½

12 Nov 1919-15 May 1920

Ordinary shale	17,191 4		716 6 0
Roman Camp shale	86,614 17		2,405 19 4¼
Curly shale	42,670 18		1,066 15 5½
Grey shale	19,679 9		266 9 10¼
Upper Grey shale	3,292 18		44 11 10
Dunnet shale	82,522 3		1,547 5 9¾
Blaes	17,713 18		73 16 2
Barracks shale	867 9		15 7 2¾
Total:	270,552 16		6,136 11 8½

16 May 1920-11 Nov 1920
- Ordinary shale — 13,004 15 — 541 17 3½
- Roman Camp shale — 87,472 5 — 2,429 15 9¾
- Curly shale — 53,489 9 — 1,337 4 9½
- Grey shale — 21,325 2 — 288 15 6½
- Upper Grey shale — 3,038 10 — 41 2 11
- Dunnet shale — 100,383 0 — 1,882 3 7½
- Blaes — 18,978 8 — 79 1 6½
- Barracks shale — 5,921 9 — 104 17 2¼
- Total: — 303,612 18 — 6,704 18 8½

12 Nov 1920-15 May 1921
- Ordinary shale — 12,868 14 — 536 3 11
- Roman Camp shale — 80,667 15 — 2,240 15 5
- Curly shale — 52,002 6 — 1,300 1 2
- Grey shale — 21,133 11 — 286 3 7¾
- Upper Grey shale — 401 5 — 5 8 8
- Dunnet shale — 82,788 18 — 1,552 5 10½
- Blaes — 16,675 13 — 69 9 7½
- Barracks shale — 13,946 16 — 246 19 3½
- Total: — 280,484 18 — 6,237 7 7¼

16 May 1921-11 Nov 1921
- Ordinary shale — 4,505 14 — 187 14 9
- Roman Camp shale — 16,921 8 — 470 0 9¼
- Curly shale — 11,338 2 — 283 9 0½
- Grey shale — 913 5 — 12 7 4
- Dunnet shale — 2,416 14 — 45 6 3
- Blaes — 1,528 0 — 6 7 4
- Barracks shale — 7,217 14 — 127 16 3¼
- Total: — 44,840 17 — 1,133 1 9

12 Nov 1921-15 May 1922
- Ordinary shale — 15,190 16 — 632 19 0
- Roman Camp shale — 58,949 4 — 1,637 9 6½
- Curly shale — 34,922 15 — 873 1 4½
- Grey shale — 2,076 13 — 28 2 5
- Dunnet shale — 11,750 15 — 220 6 6
- Blaes — 5,743 6 — 23 18 7¼
- Barracks shale — 21,818 1 — 386 7 2¾
- Total: — 150,451 10 — 3,802 4 8

Less 25% Rebate on excess Lordship over 4d per ton for year to Whitsunday £420 2s 3d

16 May 1922-11 Nov 1922
Ordinary shale	16,437 9	684 17 10½
Roman Camp shale	83,931 1	2,331 8 4
Curly shale	35,812 9	895 6 3
Grey shale	982 18	13 6 2½
Dunnet shale	50,592 18	948 12 4½
Blaes	11,822 1	49 5 2
Barracks shale	20,488 16	362 16 5½
Total:	220,067 12	5,285 12 8

12 Nov 1922-15 May 1923
Ordinary shale	14,689 12	612 1 4
Roman Camp shale	84,365 2	2,343 9 6
Curly shale	35,161 9	879 0 8¾
Grey shale	3,436 12	46 10 9
Dunnet shale	52,136 3	977 11 0½
Blaes	11,973 16	49 17 9¾
Barracks shale	17,274 18	305 18 2½
Total:	219,037 12	5,214 9 4½

Less 25% Rebate on excess Lordship over 4d per ton for year to
Whitsunday 1923:   £795 8s 5d

16 May 1923-11 Nov 1923
Ordinary Shale	13,974 17		582 5 8½
Roman Camp shale	81,345 7		2,259 11 10
Curly shale	29,005 7		725 2 8
Grey shale	3,645 16		49 7 5
Dunnet shale	52,353 17		981 12 8½
dunnet via No 5			
Roman Camp Mine	6,207 18	4½d	116 8 0
Blaes	11,631 3		48 9 3
Barracks shale	18,547 13		328 8 10½
Total:	216,711 18		5,091 6 5½

Less 25% Rebate on Lordships over 4d per ton   £369 17s 3d

12 Nov 1923-15 May 1924
Ordinary shale	14,125 11	4d	235 8 6½
Roman Camp shale	83,483 5	4d	1,391 7 9
Curly shale	4,524 15	4d	75 8 3
Dunnet shale	78,597 12	4d	1,309 19 2⅖
Dunnet via Pumpherston No 5 Mine	9,746 8	4d	162 8 9⅗
Blaes	15,106 3		62 18 10
Barracks shale	18,193 0	4d	303 4 4
Total:	223,776 14		3,540 15 8½

16 May 1924-11 Nov 1924					
Ordinary shale	12,835	16	213	18	6
Roman Camp shale	78,593	5	1,309	17	7
Curly shale	3,961	4	66	0	4
Dunnet shale	75,037	9	1,250	12	5
dunnet via No 5					
Roman Camp Mine	9,777	19	162	19	4
Blaes	14,550	16	60	12	7
Barracks shale	18,416	3	306	18	6½
Total:	213,172	12	3,370	19	3½
12 Nov 1924-15 May 1925					
Ordinary shale	13,482	14	224	14	2 4/5
Roman Camp shale	80,475	9	1,341	5	1 4/5
Curly shale	3,970	11	66	3	6 1/5
Dunnet shale	78,300	4	1,305	0	0 4/5
dunnet via No 5					
Roman Camp Mine	12,503	16	208	7	11 1/5
Blaes	11,859	11	49	8	3 3/5
Barracks shale	20,183	2	336	7	8 2/5
Total:	220,775	7	3,531	6	10 4/5
16 May 1925-11 Nov 1925					
Ordinary shale	13,408	7	223	9	5 2/5
Roman Camp shale	76,398	4	1,273	6	0 4/5
Curly shale	3,707	7	61	15	9 2/5
Dunnet shale	75,839	16	1,263	19	11 1/5
Blaes	11, 400	15	47	10	0 3/5
Barracks shale	21,569	18	359	9	11 3/5
Total:	202,324	7	3,229	11	3

The Mines were closed at Martinmas 1925

Summary:
Shale output of Broxburn Oil Company,
12 November 1877 - 11 November 1925: 20,278,678 tons 5 cwt.

Payments on output to Lord Cardross, later Earl of Buchan
£524,254 1s 11 1/20d. less Rebates of £1,585 7s 11d
plus Wayleave charges of £1,135 9s 0 2/5d on shale transported via the local mineral railways from Middleton, Stankards, Holmes, Kilpunt, Hopetoun, and Niddry, through the Buchan Estate by Broxburn Oil Company.
Total Payments to Lord Cardross/Earl of Buchan: £523,804 3s 0 9/20d

Broxburn Oil Company Wayleaves, 1906 - 1925

A wayleave is the permission for a right of way over someone else's property. A payment is made for this. Also, a fee was paid by oil companies for each ton/cwt of shale carried over the land. Shale and coal were transported via the mineral railway circuit to all the local oil works.

Nov 1906 - May 1907 - Wayleave on Middleton Hall Minerals
2850-10 @ ½d   £5 18s 9d

May 1907 - Nov 1907 - Wayleave on Middleton Hall Minerals
1752-1 @ ½d   £3 13s 0d

Nov 1907 - May 1908 - Wayleave on Middleton Hall Minerals
7707-2 @ ½d   £16 1s 1d

May 1908 - Nov 1908 - Wayleave on Middleton Hall Minerals
10190-19 @ ½d   £21 4s 7d

Nov 1908 - May 1909 - Wayleave on Middleton Hall Minerals
9909-17 @ ½d   £20 12s 11d

May 1909 - Nov 1909 - Wayleave on Middleton Hall Minerals
7249-10 @ ½d   £15 1s 11d

Nov 1909 - May 1910 - Wayleave on Middleton Hall Minerals
7478-10 @ ½d   £15 11s 6d

May 1910 - Nov 1910 - Wayleave on Middleton Hall Minerals
7871-0 @ ½d   £16 7s 10d

Nov 1910 - May 1911 - Wayleave on Middleton Hall Minerals
7109-8 @ ½d   £14 16s 1d

May 1911 - Nov 1911 - Wayleave on Middleton Hall Minerals
10226-16 @ ½d   £21 6s 1d

Nov 1911 - May 1912 - Wayleave on Middleton Hall Minerals
9829-6 @ ½d   £20 9s 5d

May 1912 - Nov 1912 - Wayleave on Middleton Hall Minerals
8198-3 @ ½d   £17 1s 6d

Nov 1912 - May 1913 - Wayleave on Middleton Hall Minerals
7119-19 @ ½d   £14 16s 9d

May 1913 - Nov 1913 - Wayleave on Middleton Hall Minerals
8213-16 @ ½d   £17 2s 4d

Nov 1913 - May 1914 -
Wayleave on 7801-3 tons from Middleton Hall Estate
@ ½d per ton.   £16 5s 0d
Wayleave on 2420-9 tons from Stankards Estate
@ ½d per ton.   £5 0s 10d

May 1914 - Nov 1914
Wayleave on 7135-8 tons from Middleton Hall Estate
@ ½d per ton.   £14 17s 4d
Wayleave on 4018-14 tons from Stankards Estate
@ ½d per ton.   £8 7s 6d

Nov 1914 - May 1915
Wayleave on 7279-1 tons from Middleton Hall Estate
@ ½d per ton.   £15 3s 3½d
Wayleave on 3834-2 tons from Stankards Estate
@ ½d per ton.   £7 19s 9d

May 1915 - Nov 1915
Wayleave on 7257-5 tons from Middleton Hall Estate
@ ½d per ton.   £15 2s 5d
Wayleave on 3434-14 tons from Stankards Estate
@ ½d per ton.   £7 3s 1d

Nov 1915 - May 1916
Wayleave on 9732-15 tons from Middleton Hall Estate
@ ½d per ton.   £20 5s 6d
Wayleave on 2834-12 tons from Stankards Estate
@ ½d per ton.   £5 18s 1d

May 1916 - Nov 1916
Wayleave on 8107-3 tons from Middleton Hall Estate
@ ½d per ton.   £16 17s 8½d
Wayleave on 2321-15 tons from Stankards Estate
@ ½d per ton.   £4 16s 8½d

Nov 1916 - May 1917
Wayleave on 6509-2 tons from Middleton Hall Estate
@ ½d per ton. £13 11s 3d
Wayleave on 1033-5 tons from Stankards Estate
@ ½d per ton. £2 3s 0d
Wayleave on 1087-7 tons from Holmes Estate
@ ½d per ton. £2 5s 3½d

May 1917 - Nov 1917
Wayleave on 5581-1 tons from Middleton Hall Estate
@ ½d per ton. £11 12s 6½d
Wayleave on 1012-18 tons from Stankards Estate
@ ½d per ton. £2 2s 2½d
Wayleave on 1591-14 tons from Holmes Estate
@ ½d per ton. £3 6s 4d

Nov 1917 - May 1918
Wayleave on 5545-1 tons from Middleton Hall Estate
@ ½d per ton. £11 11s 0½d
Wayleave on 1033-3 tons from Stankards Estate
@ ½d per ton. £2 3s 0½d
Wayleave on 1499-13 tons from Holmes Estate
@ ½d per ton. £3 2s 6d

May 1918 - Nov 1918
Wayleave on 4762-6 tons from Middleton Hall Estate
@ ½d per ton. £9 18s 5d
Wayleave on 1162-18 tons from Stankards Estate
@ ½d per ton. £2 8s 5d
Wayleave on 208-13 tons from Holmes Estate
@ ½d per ton. £0 8s 8d

Nov 1918 - May 1919
Wayleave on 4324-19 tons from Middleton Hall Estate
@ ½d per ton. £9 0s 2½d
Wayleave on 532-2 tons from Stankards Estate
@ ½d per ton. £1 2s 2d

May 1919 - Nov 1919
Wayleave on 4024-13 tons from Middleton Hall Estate
@ ½d per ton. £8 7s 8¼d
Wayleave on 54-7 tons from Stankards Estate
@ ½d per ton. £0 2s 3¼d
Excess Blaes £70 10s 6d

Nov 1919 - May 1920
Wayleave on 2572-14 tons from Middleton Hall Estate
@ ½d per ton. £5 7s 2¼d

May 1920 - Nov 1920
Wayleave on 2798-11 tons from Middleton Hall Estate
@ ½d per ton. £5 16s 7¼d

Nov 1920 - May 1921
Wayleave on 2592-14 tons from Middleton Hall Estate
@ ½d per ton. £5 8s 0¼d
Wayleave on 2685-10 tons of Niddry Shale
@ 1d per ton. £11 3s 9d

May 1921 - Nov 1921
No Wayleaves listed, so, presumably, no shale was transported.

Nov 1921 - May 1922
Wayleave on 833-10 tons from Middleton Hall Estate
@ ½d per ton. £1 14s 8¾d
Wayleave on 1662-6 tons from Holmes Estate
@ ½d per ton. £3 9s 3½d
Wayleave on 1923-2 tons of Kilpunt Shale
@ 1d per ton. £8 0s 3d

May 1922 - Nov 1922
Wayleave from Middleton Hall Shale
116-4 tons @ ½d per ton. £0 4s 10d

Wayleave from Holmes Shale 4340-18 @ ½d per ton. £9 0s 10½d
Wayleave from Kilpunt Shale 8245-8 @ 1d per ton. £34 7s 1½d
Wayleave from Hopetoun Shale 15656-2 @ ½d per ton. £32 12s 4d

Nov 1922 - May 1923
Wayleave from Holmes Shale 7549-15 @ ½d per ton. £15 14s 7d
Wayleave from Kilpunt Shale 11732-3 @ 1d per ton. £48 17s 8¼d
Wayleave from Hopetoun Shale 26385-5 @ ½d per ton. £54 19s 4½d

May 1923 - Nov 1923
Wayleave from Holmes Estate 6080-5 @ ½d per ton. £12 13s 4d
Wayleave from Kilpunt Estate 9929-19 @ 1d per ton. £41 7s 6d
Wayleave from Hopetoun Estate 20518-5 @ ½d per ton. £42 14s 11d
Wayleave from Newliston Estate 655-6 @ 1d per ton. £2 14s 7d

Nov 1923 - May 1924
Wayleave from Holmes Estate   6808-17 @ ½d per ton.   £14 3s 9d
Wayleave from Kilpunt Estate   5936-17 @ ½d per ton.   £12 7s 4½d
Wayleave from Hopetoun Estate   18504-10 @ ½d per ton.   £38 11s 0¼d
Wayleave from Newliston Estate   2491-13 @ 1d per ton.   £10 7s 8d
Wayleave from Pumpherston Estate   1234-18 @ ½d per ton.   £2 11s 5½d

May 1924 - Nov 1924
Wayleave from Holmes Estate   4755-4 @ ½d per ton.   £9 18s 1½d
Wayleave from Kilpunt Estate   5311-16 @ ½d per ton.   £11 1s 4d
Wayleave from Hopetoun Estate   16938-10 @ ½ per ton.   £35 5s 9¼d
Wayleave from Newliston Estate   1963-3 @ 1d per ton.   £8 3s 7d
Wayleave from Pumpherston Estate   2484-1 @ ½d per ton.   £5 3s 6d
Wayleave from Stankards Estate   400-10 @ ½d per ton.   £0 16s 8¼d

Nov 1924 - May 1925
Wayleave from Holmes Estate   4339-19 @ ½d per ton.   £9 0s 8½d
Wayleave from Kilpunt Estate   5329-1 @ ½d per ton.   £11 2s 0½d
Wayleave from Hopetoun Estate   17229-7 @ ½d per ton.   £35 17s 10d
Wayleave from Newliston Estate   3527-9 @ 1d per ton.   £14 14s 0d
Wayleave from Pumpherston Estate   2709-11 @ ½d per ton.   £5 12s 10⅕d
Wayleave from Stankards Estate   967-3 @ ½d per ton.   £2 0s 3½d

May 1925 - Nov 1925
Wayleave from Holmes Estate 1923-1 @ ½d per ton.   £4 0s 0½d
Wayleave from Kilpunt Estate   4745-1 @ ½d per ton.   £9 17s 8½d
Wayleave from Hopetoun Estate   12130-2 @ ½d per ton.   £25 5s 5d
Wayleave from Newliston Estate 3013-6 @ 1d per ton.   £12 11s 1³⁄₁₀d
Wayleave from Pumpherston Estate   2606-18 @ ½d per ton.   £5 8s 7⁷⁄₁₀d
Wayleave from Stankards Estate   1550-19 @ ½d per ton.   £3 4s 7½d

Shale transported to oil works via the mineral railway by Broxburn Oil Company

Wayleave charge on shale from Middleton - shale - 192, 680 tons 7 cwt	£401 7s 6¼d
Wayleave charge on shale from Stankards - shale - 26,611 tons 7 cwt	£55 8s 8d
Wayleave charge on shale from Holmes - shale - 41,847 tons 12 cwt	£87 3s 6d
Wayleave charge on shale from Kilpunt - shale - 53,153 tons 7 cwt	£177 1s 0 ¼d
Wayleave charge on shale from Hopetoun - shale - 127,362 tons 1 cwt	£265 6s 8d
Wayleave charge on shale from Niddry - shale - 2,685 tons 10 cwt	£11 3s 9d
Wayleave charge on shale from Newliston - shale - 11,650 tons 17 cwt	£48 10s 11³⁄₁₀d
Wayleave charge on shale from Pumpherston - shale - 9,035 tons 8 cwt	£18 16s 5 ⅗
Excess Blaes -	£70 10s 6d

Nov 1906 - Nov 1925
Shale carried on wayleaves by Broxburn Oil Company: 465,026 tons 9 cwt

Wayleave Payments by Broxburn Oil Company -	£1,135 9s 0 ⅖d

Broxburn Oil Company Dividends, 1877 - 1910

1877 - 1878 - A Dividend of 9%
1878 - 1879 - A Dividend of 25%
1879 - 1880 - A Dividend of 25%
1880 - 1881 - A Dividend of 25%
1881 - 1882 - A Dividend of 25%
In under five years, the shareholders got back their investment in Broxburn Oil Company. After that, it was pure profit.

1882 - 1883 - A Dividend of 25%
1883 - 1884 - A Dividend of 25%
1884 - 1885 - A Dividend of 25%

1885 - 1895 - The above Dividend has not been maintained, although dividends of 15%, 10%, and 7½% have been paid. Since the company started, £579,193 has been paid in dividends.

1895 - 1896 - A Dividend of 7½%
1896 - 1897 - A Dividend of 7½%
1897 - 1898 - A Dividend of 7½%
1898 - 1899 - A Dividend of 8½%
1899 - 1900 - A Dividend of 15%
1900 - 1901 - A Dividend of 20%
1901 - 1902 - A Dividend of 15%
1902 - 1903 - A Dividend of 15%
1903 - 1904 - A Dividend of 15%
1904 - 1905 - A Dividend of 15%
1905 - 1906 - A Dividend of 15%
1906 - 1907 - A Dividend of 15%
1907 - 1908 - A Dividend of 20%
1908 - 1909 - A Dividend of 17½%
1909 - 1910 - A Dividend of 10%

The accounts of Broxburn Oil Company Limited at 31st March 1910 showed a credit balance of £7,754 carried forward, after the payment of dividends.
The company had a reserve fund — cash in the bank — of £70,430.

The Shale Mine Years

North Greendykes Mine: 1861 to 1874

Hayscraigs Mine: 1861 to 1919

Hut Pit: 1860s to 1890

Albyn Mine: 1860s to 1905

Hut Mine: 1860s to 1912

South Greendykes Mine: 1860s to 1915

Pyothall No 5 Pit: 1860s to 1919 (Hayscraigs Pit)

Stewartfield No 1 Mine: 1860s to 1923
Stewartfield No 1 Pit: 1870s to 1914
Stewartfield No 2 Pit: 1880s to 1914
Stewartfield No 3 Pit: 1880s to 1914
Stewartfield No 4 Pit: 1880s to 1914
Dunnet Mine: c1910 to 1925

(No 1, No 2, etc, are the shaft numbers of a mine or pit.)

The Shale Bings

After the shale was processed through the oil works, the residue was tipped near the shale mines for convenience, this spent shale eventually forming large bings, which were of a reddish colour. The word bing comes from the Old Norse word *bingr*, meaning a heap or pile (of waste).

These grew up near Greendykes and Stewartfield. Their approximate size was:

Albyn - Height: about 150 feet.
Weight: around 2,000,000 tons.

Greendykes - Height: about 280 feet.
Weight: around 5,000,000 tons.

Stewartfield - Height: about 200 feet.
Weight: around 3,000,000 tons.

Much of the shale from the bings has been removed for use as foundation in road building which took place from the 1960s to the 1990s.

In recent years, using the shale in this manner seems to have ground to a halt.

Local Names for the Shale

Some of the local shale names are rather confusing, because local companies used their own names for the seams, and these were often different from the names used in the Mining Authority Records for the same shale seams.

Further confusion is sometimes caused by the use of two shale names which suggest the same shale. Camps shale and Roman Camp shale are listed by Broxburn Oil Company Ltd, and the names suggest that they should be one and the same, yet this does not seem to be so.

Figures listed for shale-depths can bear little comparison to the depth at which the shale was actually found, and can be way off the mark. They are the levels at which the various shale was originally deposited.

Shale depths were by no means regular or consistent. A particular shale might be found at different depths, even in the same locality.

Deposits of shale were laid down many millions of years ago, and sometimes underground pressure broke seams and pushed sections of them hundreds of feet towards the surface, or dragged them down to the depths, e.g the Dunnet Shale at Broxburn was elevated to within 720 feet of the surface, while at Hayscraigs shale was thrust right to the surface because of extreme forces at work below ground in the distant past.

Shale miners working a seam might come upon a sudden break in the shale, the rest of the seam apparently having vanished – perhaps cast down into the bowels of the earth, never to be found again.

Shale Worked by Broxburn Oil Company Limited

Opencast Shale
Obtained by open excavation - a quarry.
(Hayscraigs)

Blaes (Shale containing no oil-bearing material)

Raeburn Shale around 255 ft
(Upper Shale)
(Dam Shale)

Upper Grey Shale around 355 ft.
(Mungle Shale)

Grey Shale around 630 ft

Fells Shale around 1,080 ft

Broxburn Shale around 1,250 ft
(Ordinary Shale)

Dunnet Shale around 1,900 ft

Barracks Shale around 2,000 ft
(Under Dunnet Shale)
(New Shale)

Camps Shale around 2,390 ft

Roman Camp Shale around 3,000 ft

Curly Shale around 3,260 ft
(No 3 Shale)
Shale with an undulated, rippled appearance, due to pressure.

It has been estimated that about 75% of the shale still lies underground.

Glossary

admonished - warned regarding his conduct.
afterdamp - gas left after an explosion containing a large quantity of carbon monoxide.
amain - violently, at full speed.
bench - landing place.
bottomer - person who loads or unloads hutches at the bottom of a shaft.
braker - the man in charge of the winding-engine which raised the shale in a shaft.
cage - the carriage or platform used to raise men and material in a shaft.
cera-lamps - lamps with a wax taper.
char - solid material left after gases and tar have been removed from a carbon based material.
crowns - set of timbers.
cuddie-brae - an inclined roadway on which a bogie (cuddie) is used to counter-balance the weight of the hutches.
cut chain - a chain with links where hutches could be attached on an incline at intermediate points.
deferred shares - shares which yielded large dividend payouts. After all other classes of shareholders (those holding ordinary shares or preference shares) had been paid out, the holder of deferred shares had access to all remaining profits.
dook – road on an incline.
drawer - person who takes the shale (or waste) from the working-face to the haulage road.
dry - sudden fracture.
engine plane - a roadway on which hutches are pulled by an engine.
firedamp - methane gas. Highly inflammable, and forms an explosive mixture with air.
Gay-Lussac tower - prevented dangerous gas emissions. Stopped nitrogen oxide gas from escaping into the atmosphere. The fumes were absorbed and sent back to the Glover tower where they were re-used.

Glover tower - a tower about 30 feet high and 10 feet wide, made of sheet lead. Acid from the chambers was collected within this. Sulphurous acid was converted to sulphuric acid.
holding company - a company formed to hold the shares of other companies.
holing - removal of material above or below a seam by cutting or blasting.
in-going-eye mine - a mine starting from the surface of the ground.
inter alia - among other things.
Kessler plant - the Kessler furnace concentrated the sulphuric acid, giving it the highest strength possible.
Lordship - the territory of a Lord.
Lordship payments - charges for minerals extracted from his property.
lypes - irregular faults in the strata of a shale seam.
marine sperm - marine sperm was whale oil from the sperm whale. It was not a true oil, more of a liquid wax. The oil produced at the oil works must have been classed as its equivalent. So, a quality lubricating oil.
oncostmen - mine workers paid by the day.
overman - person in charge of underground operations.
parent company - one which held the majority (or all) of the shares of another company.
parting - the separating of the shale seam from its roof.
petroline - a type of paraffin oil.
pinching - using a crowbar.
place - a length of face assigned to each miner.
rake of tube - a set of hutches coupled together.
roadsman - underground official who looks after roads and rails.
room - working place in area where stoops (pillars) of shale have been left to support the roof (stoop-and-room workings).
screens - parallel bars over which the material is passed to free it from dross (small pieces which pass through the screen).
spragged - not spragged - no pit props.
stooping - leaving pillars of shale to support the roof.
stooped waste - workings where the stoops (pillars of shale left to support the roof) have been worked out.

top shale - the highest of two or more divisions of a shale seam.
trapper - boy who opens and shuts a trap-door, directing the flow of air for ventilation purposes.
trimmer - someone who arranges the material in wagons while they are being loaded.
wayleaves - rights of way rented to another.
wayleave payments - charges for minerals from elsewhere passing through the owner's lands.

Index

A

Abadan Oil refinery 79
Albyn Oil Works
    mentioned 20, 22,23, 24,, 25, 26,31,45, 53, 57, 90, 105, 106, 107, 108, 109
    closure 81
Albyn shale bing 133
Albyn Shale Mine 16, 43, 132
Almondell House 9, 19, 20
Almondfield 20
ammonia 14, 38, 39, 40, 41, 92, 100, 101
ammonium nitrate 9, 14, 55, 75
Anglo-Iranian Oil Company 80
Anglo-Persian Oil Company
    mentioned 45, 49, 52, 75, 77, 78, 84, 86
    history 79-80
Arches, the 20, 22, 23, 104, 105, 106

B

Barracks shale 135
Bell, Robert
    mentioned 10, 11, 17, 19, 20, 21, 22, 23, 24, 25, 26, 28, 29, 30, 31
    Albyn Oil works 25, 26
    Almondell House 9
    Broxburn Lodge 7
    Broxburn Oil Company (Ltd) 25
    Broxburn Oil Works (Refinery) 25, 26
    Broxburn Shale Oil Company (Ltd) 6
    Chairman 31, 110
    Clifton Hall 9
    death 11
    Eglinton Crescent 9
    Fever Hospital 10,11
    first shale mine 3
    Glasgow Oil Company Broxburn (Ltd) 20
    Grosvenor Street 25
    Holygate 4
    lease 1
    marriage 7
    North Greendykes Shale Mine 4
    oil works 17, 20, 22, 23, 24, 25, 104-108
    Public Hall 10
    refinery 23, 107,108
    Wishaw 4
    Wishaw Iron Works 4
Black, Sir Frederick William 52
Blaes 135
BP 29, 80
Brown, William 6
Broxburn Candle Works (House) 33, 64, 83
Broxburn Colliery 7
Broxburn Gas Works 31
Broxburn Lodge 7, 8, 20, 90
Broxburn Oil Company (Ltd)
    mentioned 9, 43, 48, 49, 53, 75, 76, 77, 82, 83, 86, 90,92, 99, 130, 131, 132, 134, 135
    agreement with Robert Bell 28, 29
    closure 81-82
    company document 27, 48
    Court of Inquiry 76-77
    dividends 131
    Henderson, Norman Mcfarlane 90
    incorporation 25

management 110
mines to Dunnet Shale 93-98
oil works 26, 108-109
registered office 33
Roman Camp 35, 84, 85
Scottish Oil Agency 47
shale output 110-124
shale payments 110-124
strike at oil works 33, 53, 55
wayleaves 125-130
workforce 110
Broxburn Oil Works (Refinery)
26, 46, 53, 54, 57, 61, 63, 64,
66, 77, 78, 83, 85, 86, 90,
108, 109
closure 81, 82
description of 99-103
dismantled 83, 86
Broxburn Refinery (Oil Works)
31, 35
Broxburn retort
Broxburn Shale 4, 6, 16, 17, 20,
22, 28, 55, 94, 104, 135
Broxburn Shale Oil Company
(Ltd) 4, 17, 20, 104
shareholders 6
Buchan, Earl of 1, 28, 30
payments to 113-124
Buchan Estate 13
Buchan Oil Works 20, 22, 23, 24,
105, 106
Burmah Oil Company 79

C

Camps Shale
Cardross, Lord — see Buchan,
Earl of
Carledubs mines 54
Churchill, Winston 79, 84, 86
Clark, William 93, 110

Clifton Hall 9, 10, 11
Conner, Benjamin 21
Cooke, Henry Rowland 52
Crawford filling apparatus 103
Cunningham, John 1
Curly Shale 16, 43, 135

D

Dalrymple, Agnes 7
Dam Shale 135
D'Arcy, Wiliam Knox 79
Deutsche Bank 80
Dick, Alexander 6
Dixon, James Stedman 110
Drumshoreland 41, 61
Dunnet Shale 93, 94, 96, 98, 134,
135
Dunnet Shale Mine 16, 49, 53,
54, 55, 93, 94, 96, 98, 132
closure 81, 82

E

East Mains 25, 26, 31, 108
Eglinton Crescent 9, 11
Erskine, David Stuart 30
Erskine, Shipley Gordon Stuart 30
Europaische Petroleum Union 80
Excise Duty relief 46, 54, 84, 86

F

Faulds, Robert 6, 17, 21
Fatal Accidents
Anderson, Robert 34
Anderson, Robert 44
Armstrong, William 32
Beith, Andrew 33
Bell, William 34
Beveridge, David 25

Bisset, David 32
Black, Henry 31
Bonnar, James 35
Brogan, Michael 43
Bryans, J. 25
Crawford, William 45
Crawford, William 53
Cushley, John 46
Danks, Francis 32
Ellis, Joseph 36
Ferguson, David 36
Given, David 55
Imrie, John 32
Kelly, Michael 33
Kennedy, Robert, 45
McCord, Hugh 43
McLauchlan, William 37
McNee, James 36
McVey, John 44
Millar, John 46
Miller, Robert 54
Munro, Alexander, 49
Murray, Daniel 37
Neil, John 32
O'Hare, Richard 54
Potter, James 53
Practice, Robert 43
Shanks, George 44
Shields, John 22
Sibbald, George 55
Sneddon, Richard 33
Steven, John 32
Stewart, John 35
Fells Shale 135
Ferguson, George 29
Fernie, Ebenezer Waugh 19, 20, 104
Fernie, Edward 17, 19, 20, 104
Fever Hospital 10,11
Flintshire Oil and Cannel Company 17
Forbes, D. 29

Fraser, William 47, 52
Furnival's Inn 19

G

Garrow, Duncan 52
General Strike 81
Glasgow Oil Company Broxburn (Ltd) 20, 21, 22, 23, 24, 105, 106, 107, 108
  shareholders 21
Greendykes 17, 20, 22, 23, 24, 25, 26, 31, 104
Greendykes Farm 7
Greendykes North Pit (North Greendykes Shale Mine) 4, 16, 24, 132
Greendykes shale bing 133
Greendykes South Pit (South Greendykes Shale Mine) 16, 24, 46, 132
Greenway, Sir Charles 52, 84
Grey Shale 135

H

Hamilton Advertiser 22
Hamilton, James 21
Hayscraigs 1, 2, 3, 7, 49, 134, 135
Hayscraigs Farm 1
Hayscraigs Shale Mine 16, 32, 34, 49, 58, 132
Hayscraigs Shale Pit (Pyothall No 5 Pit) 16, 25, 49, 132
Henderson, Norman Macfarlane 31, 34, 38, 39, 40, 41, 42, 90, 92, 100, 101, 102, 110
  history of 90-91
  death 92
Henderson retort 34, 90, 92
Henshaw, William 6

Holmes Estate, 7, 127, 128, 129
Holmes Oil Company 7
Holmes Oil Works 7
Holygate 4, 5, 20, 31, 61
Holygate Farm, 4, 5
Hopetoun Estate 128, 129
Hurst, William 21
Hutchison, Thomas 17, 19, 20, 22, 23, 104, 105
Hut Shale Mine 16, 132
Hut Shale Pit 16, 132

K

Kennedy, William 28, 29, 30, 110
kerogen 12, 14
kerosene 14, 23
Kilpunt Estate 128, 129
King, John Nicholls 28, 29, 30
Kirkhill 28

L

Liddell, James 23, 24, 25, 106, 107
Liddell, James & Company 108
Lighthouse oil 102
Lloyd, John Buck 52
Love, William 99, 100
lubricating oil 14,
Lythgoe, James 21

M

Maitland, Sir James Gibson 9
Maitland Street 25
Marine sperm oil 102
Martinmas 55
McLellan, Walter 21
Middleton Hall 52
Middleton Hall Estate 125, 126, 127, 128

Miller, James 20, 21, 22, 23, 24, 25, 105, 106, 108
Mitchell, William 29
Mungle Shale 135
Murray, Robert Alexander 52
Murray, W. 110

N

naphtha 14, 40, 101, 102
Newbigging Brick Works 7
Newbridge 9
Newhouses Road 10
Newliston Estate 128, 129, 130
Newliston Shale Mine 20
New Shale 135
Nichols, Hubert Edward 52
Niddry 128, 130
North Greendykes Farm 3
North Greendykes Shale Mine (Greendykes North Pit) 4, 16, 24, 132

O

Oakbank Oil Company 47, 49, 75, 76, 77
Oakbank Oil Works 90
Opencast Shale 4, 16, 111, 135
Ordinary Shale 112-124, 135

P

paraffin 13, 14, 17, 23, 40, 41, 64, 75, 92, 101, 102, 137
petroleum jelly 14, 55
Petroline 102
Philpstoun Oil Works 47, 49
Poole Hall 17, 18
Poynter, John Edgar 20, 22, 23, 24, 105, 106

Pumpherston estate 129, 130
Pumpherston Oil Company 47, 49, 75, 76, 77
Pumpherston Oil Works 82
Pumpherston Refinery 85
Pyothall 1, 2, 3
Pyothall No 5 Pit (Hayscraigs Shale Pit) 16, 25, 49, 132

R

Raeburn Shale 135
rail-wagon repairs 83
refining process 40-42
retorting process 38-39
Richmond, Sir David 110
Robertson, Alexander 6
Roman Camp 35, 41, 61, 134
Roman Camp Oil Works 82, 84, 85, 99, 100
Roman Camp Shale 112-124, 134, 135
Ross, James and Company 47, 49, 75, 76, 77
Rothschilds 79

S

Scotsman, The 34
Scott, Alex 7
Scott, William 21
Scottish Oil Agency 47, 48, 49, 53, 75, 76, 77
   company document 47
Scottish Oils 47, 49, 52, 53, 55, 75, 76, 77, 78, 84, 85, 86, 109
   company document 50, 51
   Court of Inquiry 55, 76-77
   Royal Navy contract 84
   shareholders 52
Shah of Persia 79

shale bings 132
shale geology 12
shale oil 7, 12, 13, 14, 19, 20, 25, 55, 64, 72, 83, 84
Shandwick Place 25
Shaw, Sir Archibald McInnes 110
Shotts Iron Company 90
Simpson, George 25, 26, 31, 108
South Greendykes Shale Mine (Greendykes South Pit) 16, 24, 46, 132
Spylaw House 17, 18
Stankards Estate 126, 127, 129
St David's Oil Works 17
Steel, Sir James 110
Steele, Dr. James 17, 19, 20, 22, 23, 24, 104, 105
Stephen, William 6
Steuart, D. R. 42
Stewart, John Douglas 52
Stewartfield 17, 19, 20, 22, 23, 24, 25, 31, 43, 46, 54, 55, 68, 71, 73, 104, 133
Stewartfield Farm 17
Stewartfield shale bing 133
Stewartfield Shale Mine 16, 34, 53, 132
Stewartfield Shale Pits 16, 43, 132
still-coke 101
Strathbrock 28
Strathbrock Centre 10
strike at Broxburn Oil Works 33, 53, 55
sulphuric acid 14, 39, 40, 55, 101, 137
Sulphuric Acid Plant 35, 83

T

Thomson's Coal Pit 1

U

Under Dunnet Shale 135
Union Canal 64, 73
Uphall 7, 10, 52, 61, 82, 92
Upper Grey Shale 135
Upper Shale 135

V

Vallance, James 20, 22, 23, 104, 105, 106

W

Waddell, John 28, 29, 30
Watson, Robert Irving 52
wax 13, 14, 41, 55, 64, 92, 102, 136, 137
Wayleaves 124, 125-130
West Lothian Courier 38, 49, 54, 81, 82
Wishaw 4, 17, 20, 24
World War I 45, 53, 80

Y

Young and Meldrum's Bathgate Works 90
Young's Paraffin Light and Mineral Oil Company 47, 49, 75, 76, 77

www.ingramcontent.com/pod-product-compliance
Lightning Source LLC
Chambersburg PA
CBHW070808100426
42742CB00012B/2300